The Circular Advantage

How Artificial Intelligence and Reuse Are Redefining Profitability and Sustainability

By Justin and Ryan Andrews

Brothers and Co-founders of REUZEit

Foreword by

Andrei Klimanov

Acknowledgments

No book like this is ever the product of just one or two voices.

We are deeply grateful to the people who helped to make this book a reality, through dialogue and obstacles and hands-on involvement and their support for a better future.

We appreciate our colleagues, partners, and team members at REUZEit for their willingness to participate in unclear situations. You've taught us that building circularity takes more than just vision; it requires hustle, muscle, and a whole lot of tenacity. We learned that innovation happens in warehouses, loading docks, labs, and spreadsheets just as much as it does in conference rooms.

To the technicians, operators, and logistics pros across the world: your hands, judgment, and dedication are the real engines of the circular economy. You don't get enough credit, and we hope this book plays a small role in changing that.

We stand on the shoulders of thought leaders and practitioners who came before us, especially the work of the Ellen MacArthur Foundation, William McDonough, and institutes that measure and report circularity metrics, like Circle Economy and PACE (Platform for Accelerating the Circular Economy). These leaders established a regenerative future as a viable possibility when most people considered it abstract. They made it thinkable. We tried to make it practical.

We appreciate our families for providing the necessary time along with support and patience needed to create enduring structures—including businesses and systems and books.

And finally, to the reader: you deserve our deepest gratitude for advancing this conversation. Whether you're redesigning products, implementing circular logistics, writing policy, or just rethinking how your organization uses resources—you're the continuation of this story.

Keep going.

— Justin and Ryan Andrews

Table of Contents

Foreword

I can recall, on more than one occasion, standing in a disorganized warehouse where rows of dormant equipment that once were considered the pinnacle of industry innovation sat discarded and unused. We tend to neglect the worth of our current possessions as we chase forward progress. And this behavior isn't limited to equipment maintenance. It encompasses humanity's approach to all resources.

Throughout my twenty years in software engineering and data science I have observed the fast-paced transformation of the technology industry. Theoretical advancements often fell apart when attempts were made to translate them into practical solutions. People praise new discoveries, yet implementation is often the hardest step.

That's why this book is so valuable. It moves from strong theoretical underpinnings to explore practical implementation methods. It therefore serves as an actionable manual for those who want to transition from discussion to meaningful application.

The Need for Circularity Continues to Grow

The world stands at an inflection point. Supply chains are fragile. Energy grids are strained. Multiple elements that humanity once considered plentiful—such as copper, fossil fuels, and clean water—are now so overdrawn that they are central to international political conflicts. Our routine of extraction and consumption and waste continues, often without acknowledging any changes in our circumstances.

Circularity must soon function beyond its status as a buzzword (and a lesser-known one, at that). It's a survival strategy.

The concept of circularity is often misunderstood. The general public tends to associate this concept with a recycling bin and environmentally friendly product labels. They think it's something for environmentalists and political analysts to focus on. Practical applications, particularly in the world of industry, remain largely absent. And businesses continue to transform

2

our environment with a rigorous eye toward profit and production.

This is not a manifesto. It's a manual. In this book, brothers Justin and Ryan Andrews create pathways toward building the world they envision, instead of just depicting the world as it is, with all its many problems. It is distinct from other publications on this topic because of its intentionality in connecting ideas and solutions to both the boardroom and the factory floor.

Why Justin and Ryan Asked Me to Write This Foreword

Throughout my career, I've worked at the intersection of innovative ideas and operational challenges. When I joined REUZEit as Chief Technology Officer, I became part of an organization that operated from a solid foundation of logistical systems and international compliance requirements, as well as strong asset management in warehouse environments that all had time, budget, and environmental constraints. REUZEit was an ideal environment for a case study on the implementation of Vertical AI.

My systems development experience has spanned various sectors, including advanced energy systems in Europe and telecom and infrastructure platforms in California. No matter the setting, failure always had costly consequences. Over time, I've learned that sophisticated concepts must pair with practical knowledge of specific situations to survive in real-world applications. And some of the most important practical knowledge can be derived from data.

The circular economy attracted me because of its tangible, reachable goal: to improve supply chain management while conserving assets and being smarter with our use of existing resources. We understand the issues and, in many cases, there are known solutions. Combining existing data with business logic will enable our ability to succeed. And that's where I've situated my work in the field.

A Guide to Circularity from Brothers Who Built a Business Around It

I believe in this book. It's one of the few works in the sustainability space written by people who have done the work. Justin and Ryan understand the tension between legacy systems and bold innovation. They've built workflows, not just whitepapers. They have experienced firsthand both the challenges and incredible benefits of the circular economy. They have worked with procurement staff and engineers. They've handled worldwide surplus logistics while learning firsthand what happens when Fortune 500 organizations implement reuse programs. They are not system theory writers; they are entrepreneurs who have redesigned systems in real-time for businesses that have needed to keep an eye on their P&L and keep shareholders happy.

In an era where everyone has an opinion on AI and LinkedIn posts about "going green" abound, that level of earned authority matters.

Let's be honest: most books about sustainability are either extended philosophical essays or essentially long tech marketing brochures.

This one is neither.

What you're holding is a grounded, no-nonsense guide that doesn't sugarcoat the challenges of the circular economy—but it doesn't get lost in academic jargon, either. It speaks the language of operations, supply chains, risk management, and margin pressure. The authors understand the reality most leaders face: You can't just "go green" by wishing it so. You need systems, tools, incentives, and—above all—results.

I was struck by how much this book respects both the theory and the execution. You'll find ideas rooted in regenerative design and industrial ecology, yes. But you'll also find spreadsheet-level tactics. You'll learn how to manage surplus assets, how to avoid sunk costs, and how to design workflows that reuse *before* they recycle.

The authors don't claim to have invented circularity. What they've done is more valuable: they've operationalized it.

The Andrews brothers have also taken a clear-eyed view of AI, looking at it not as a magic wand, but as a tool. This book explains how, when Vertical AI is purpose-built for surplus management, it can streamline complex decisions such as identifying equipment, predicting value, and reducing waste. This is the kind of integration that will move us all forward, in which AI augments human expertise, not replaces it.

Crucially, *The Circular Advantage* never loses sight of the bigger picture in that circularity isn't just about waste reduction—it's about resilience. A transition to the circular economy is about building companies, cities, and systems that can adapt, recover, and regenerate.

It's not a pitch deck. It's a blueprint.

Vertical AI: The Hidden Backbone of Circularity

When people talk about AI, they're usually thinking of chatbots. But the most transformative AI—Vertical AI— isn't designed to talk. It's designed to work.

Rather than participating in general conversations or summarizing search results, Vertical AI focuses on developing intelligence for solving industry-specific problems. Vertical AI is built for specificity. For nuance. For edge cases that don't show up in glossy product demos but make or break a real-world operation. And nowhere is that specificity more necessary than in the circular economy.

Circular systems depend on context. Knowing when a piece of equipment can be reused, or if it's time to recycle. Understanding the difference between a cracked display and a critical hardware failure. Estimating depreciation, condition, logistics cost, and downstream utility—not in a vacuum, but in real time. That's not general intelligence. That's domain intelligence.

The emergence of Vertical AI presents an ideal opportunity for the widespread adoption of circularity, right now. The authors of this book don't just recognize that—they've built for it.

Through the lens of Vertical AI, Justin and Ryan show how data—from asset images to service histories to warehouse processing times—can be structured into something actionable. Deploying intelligent systems built from proprietary data sets can enable business owners to do things like reduce waste by flagging reuse opportunities before equipment hits the landfill.

In my experience, flashy foundation models don't matter nearly as much as the ingestion pipeline. You don't need a billion parameters to solve a painkiller problem. You need the right data, the right workflows, and the right humans in the loop. The future combines human and machine elements into a continuous feedback mechanism.

This book delivers exactly what readers need: guidance on functional AI integration rather than hyped-up, and often empty, AI promises. Numerous AI thought leaders today expound theoretical concepts about automation and optimization while remaining distant from the practical challenges of implementation. The Andrews brothers lie on the other end of this spectrum. Through a clear explanation of processes built from their own experiences, the authors present the inevitable AI deployment roadmap that sectors like logistics and manufacturing and healthcare and education will adopt in the future.

The Next Step Is Yours

The conversation about sustainability continues to

dominate our discourse, yet many of us feel over-whelmed about the size of the problem and therefore don't take action. At the same time, our social media feeds are filled with AI-related news, but we have not made sufficient progress in transforming it into practical value at a societal level.

This book does not provide a ready-made, one-size-fits all solution to these challenges. However, the authors demonstrate how AI can be used to quickly scale systems that will make a measurable difference in sustainability measures, especially in the realm of industry.

Justin and Ryan acknowledge that transitioning to circularity requires time, testing, education, planning, buy-in, and more, all while stressing that AI alone cannot solve all problems. That said, this book also makes clear the need for dedicating yourself to this work as well as provides effective pathways in achieving, and benefiting from, a circular business model and its superior systems. The circular model operates as a competitive advantage rather than an added expense.

Reading these pages will provide you with fresh concepts and useful case studies from the circular economy. What's more: This book will serve as a catalyst to start important internal discussions within your organization. You will identify business opportunities that have caused organizational inefficiencies for too many years. You will have the foundation to analyze existing systems that need change. You will come up with countless questions and ideas for your team as well as your vendors and board members.

This book will help you move beyond traditional volume metrics and embrace environmental impact reduction, for the benefit of both your organization and your personal growth.

Our world requires action over additional conceptual ideas. We need execution. *The Circular Advantage* can serve as a guide toward thoughtful action.

The following step belongs to you.

Introduction

The day I brought home my first printer, lugging the boxy beast through the front door of my studio apartment, is as clear a memory as bringing home a new pet. I found the perfect place for it, plugged it in, and fed it paper and ink. I may have even pet it.

Indeed, my early days with this cutting-edge marvel of modern technology felt like a honeymoon. I can still smell the strangely appealing scent of fresh ink on hot, clean paper. I can hear the machine working—*click, SCREEEEECH, click*—to print everything from my final papers to my class schedule, these hardcopies the tangible proof of my diligent work. Reflecting on that transition time in my life, it's as if the printer helped carry me across the threshold from university student to legitimate professional! I even printed out images I'd downloaded from the brand-new *world wide web*, pin-striped CMYK photos that accompanied the mural of magazine cut-outs plastering the walls of my apartment. Looking back, it was wall art that could only have existed in the '90s home of a college kid.

But over time, inevitably, the printer's charm wore off. The dreaded paper jams were occasional at first—a single crumpled piece of paper stuck inside the machine's jaws, and me doing careful surgery to extract it—but they quickly became a common annoyance. Then, there was the ink. Ah, the ink. The printer drank it up the way a dog gulps water after a long walk on a hot day, the big difference being that printer ink doesn't come from a tap. Refilling those cartridges was like purchasing liquid gold.

Two years in, the printer and I had, at best, a strained relationship. It sputtered and whined like an old car on its last legs. Even the simplest print jobs took ages, and the print quality had degraded from charmingly pinstriped to barely legible. Finally, it refused to print anything at all, flashing an error message I couldn't understand, let alone correct. Troubleshooting guides and early internet forums were no help. I admitted defeat. In a precursor to that infamous scene from *Office Space,* and dramatic contrast to the enthusiasm with which I'd once embraced the new marvel of a device, I threw it from my balcony. It sat in the grass for a few days, like a corporate art installment in my backyard. And then, I picked it up and heaved it into the dumpster, where it waited again for removal by the big gas-guzzling truck that would drop it off in a landfill. The morning I heard the garbage truck drive up, I couldn't help but feel guilty: this once-useful machine, this equipment I'd so doted on, was now just another piece of waste.

After the garbage pickup, I took a long walk. And coming back into my apartment, I saw my things—all the many gadgets and devices that I'd amassed—with new eyes. I realized that each of these, too, would likely one day stop working or become obsolete. How quickly we go through products, I thought, each one requiring such resources to manufacture, only for them to end up in a huge pile of other defunct and useless goods in a part of town that most of us never visit. (Isn't it convenient, how we situate landfills where nobody has to see them? How the phrase *throwing something "away"* lets us push that thing out of our minds, as if it no longer exists?) It dawned on me then, as it has on so many of us in recent years, that this constant cycle of production and disposal is completely unsustainable.

Manufacturing new things requires massive amounts of raw materials and energy, including those derived from processing ore and oil. And it's not just the manu-facturing process that draws on these finite resources, either. Transporting, assembling, packaging, retailing, disposing, and even traditional recycling all depletes our natural resources and contributes to worldwide pol-lution. This grossly inefficient consumption cycle in-cludes everything from exotic plastic extrusion to the burning of fossil fuels to harmful component disposal. *To put it plainly, each new device we buy increases the world's environmental load.*

My first printer taught me more than just how frustrating paper jams and expensive ink cartridge replacements can be. It opened my eyes to the bigger implications of our consumption patterns. As my awareness—and concern—grew, I began to research the world of refur-bished and repaired electronics. On a personal level, I developed my own style of saving old gadgets from landfills, simultaneously extending their usefulness and making them more affordable. On a professional level—well, that's the bigger story we'll explore to-gether. In short, this hands-on work has made me su-premely aware of the product lifecycle and deeply committed to making environmentally friendly purchasing and consumption decisions. I'm here to help you and your companies do the same.

Who We Are: The Andrews Brothers

My name is Justin Andrews, and my brother Ryan and I have dedicated our careers to negotiating the complexities of introducing sustainability into corporate models. As the co-founders and co-CEOs of REUZEit, we've helped some of the world's largest companies recover millions of dollars in excess capital investments while preventing mountains of used equipment from filling landfills worldwide. We've been around for over 15 years and serve leading names in the biotech and pharmaceutical industry. In a nutshell, our comprehensive services empower companies to recoup capital investments by reusing assets at their own organization, by reselling them through our online network, or via project cost avoidance. At REUZEit, we call this *circular equipment management.* This work allows our partners to manage their excess assets at zero net cost.

Our goal is to share our expertise, hard-earned through years of hands-on experience and research, with you via this straightforward guide to the circular economy. We'll draw on our personal experiences—including everything from amusing anecdotal evidence to wildly impactful multi-year case studies—to help you understand the value of this work as well as how to implement it. This isn't a philosophical treatise. We're not academics. We're business owners. And we're going to provide you with actionable insights to put sustainable theories, like that of the circular economy, into practice. Justin sitting at home with his printer guilt, which spurred an individual consumption change, matters. But what we're most excited about is scaling this mindset, and bringing it to the desks and the strategy plans of the biggest offenders: corporations.

First up: why applying sustainable practices in the corporate world makes sense—for both the future of our environment and your bottom line.

Environmental Responsibility: A Smart Business Decision

"When sustainability is viewed as being a matter of survival for your business, I believe you can create massive change."

In 2020, researchers at the University of Trento in Italy discovered an intriguing trend[1]: based on data from Italian innovation startups between 2009 and 2018, **environmentally-focused firms are twice as likely to survive the early stages of development than their less-eco-conscious counterparts.** Startups committed to sustainable practices and green concepts are not only benefiting the environment but are also increasing their chances of success, laying a solid foundation for their own long-term viability in an increasingly competitive marketplace.

What is behind this intersection of business performance and environmental responsibility? First and foremost: consumers. Today's consumers are drawn to companies that demonstrate a commitment to sustainability, both as new and repeat customers.

According to a 2024 NYU Stern report[2], products that are marketed as sustainable have continued to grow in shares each year since 2013, even in the years of the COVID pandemic and subsequently high inflation.

Tapping into the expanding market demand for eco-friendly products and services offers entrepreneurs a devoted client base *from the get-go* (i.e., the point of launch)—one that is willing to pay a premium for products from organizations that share their beliefs. This, in turn, can make it easier for burgeoning companies to access financing from organizations and individuals eager to be associated with green initiatives.

Building on these consumer preference trends, environmental consciousness in messaging promotes a positive—and therefore profitable—brand image among your existing customer base. In 2023, McKinsey found[3] that brands with more sustainable products in their lineup enjoy more *loyalty* from customers, with 32 to 34% of customers returning for a repeat purchase three or more times a year from companies "that garner more than half of their sales from products making ESG-related claims. Repeat rates were under 30% for brands that received less than half of their sales from products with ESG claims. This difference is small as far as percentage points go, and it's hard to capture the monetary value of a repeat customer. However, as any seasoned marketer will tell you, it costs much less to keep a customer than to convert a new lead or prospect. Brand loyalty pays dividends—and so does *referral business.* Hoping that your customers will recommend your product or service to their friends and family members? In 2022, PwC found[4] that 48% of people often or always take a company's environmental record into account when deciding whether to do so.

Even more tangibly, sustainable techniques often result in in-house operational efficiencies that offer a competitive advantage in the market. Avoiding waste and optimizing resource use can cut down on both ongoing costs and long-term expenditures, acting as a financial buffer that less environmentally conscientious peers may lack. **Less waste → decreased costs → increased profit margins.** What business owner, or manager or director or executive, doesn't want that? Similarly, when it comes to heavy machinery in particular, efforts made toward minimizing run times will both cut down on their energy use *and* wear and tear, meaning your capital equipment will likely be able to serve your business needs for longer.

On the flipside, if you're more motivated by the downsides of inaction, let us put it this way: **companies that *fail* to adopt sustainable practices will lose their**

competitive edge. According to a 2023 Thomson Reuters report[5], 71% of C-suite executives and functional leaders anticipate the growing significance of environmental, social & governance [ESG] factors in corporate performance. That is, they understand and are preparing for ways in which the integration of ESG increasingly drives financial performance and risk management, beyond pure compliance requirements. It's clear that a big focus of leading companies today is on building sustainable, resilient models that can withstand economic disruptions and foster long-term growth.

In summary, other risks inherent to not embracing sustainability practices include the following[6]:

1. **Increased costs:** Unsustainable practices may lead to inefficiencies and wasteful resource management, ultimately (and avoidably) increasing operational costs that strain an organization's financial viability.

2. **Limited access to capital and investment:** Investors and financial institutions are more and more often considering sustainability criteria when making investment decisions. Companies with unsustainable practices may find it

challenging to attract investment or secure financing for growth.

3. **Legal and regulatory risks:** Ignoring largescale sustainability can expose businesses to legal and regulatory risks. Specifically, non-compliance with environmental or social regulations can result in costly fines, penalties, or even legal actions that disrupt business operations. Want a real-world example? A Pew Research study conducted in late 2024[7] found that 79% of American adults favor providing a tax credit to businesses for developing carbon capture/storage—and 68% favor taxing corporations based on their carbon emissions. These numbers hold across party lines.

4. **Supply chain disruptions:** If sustainability is not considered throughout the supply chain, businesses may face disruptions in the availability of resources or materials critical to their daily operations. This can affect production and cause delays in delivering products or services to customers.

5. **Missed business opportunities**: Not embracing sustainability can result in missed opportunities for innovation and new market entry.

Sustainable practices often lead to new avenues for growth and diversification.

6. **Reputation damage:** Not prioritizing sustainability can lead to negative publicity. Being associated with unsustainable practices, or those that actively damage the environment, can result in loss of trust and credibility on the part of increasingly environmentally conscious consumers and stakeholders.

7. **Weakened business relationships:** Similarly, a lack of commitment to sustainability can strain partner relationships. If one partner perceives the other as not sharing their same values or goals, it may lead to conflicts and eventually break down the collaboration.

There is always a risk of failure in starting, running, or making a business change. However, *not* taking steps can also lead to failures—often costly failures that could have been prevented. When it comes to sustainability, the cost of inaction is real; businesses that neglect the matter will face challenges adapting to quickly evolving market demands and stricter regulations. But on the flipside, businesses that *embrace* sustainability—including making changes that may feel painful in the short term—will thrive even more than we are yet prepared to understand or anticipate.

Put simply, in today's market, sustainable practices are crucial for ensuring a business's resilience, and therefore its long-term viability. And one of the most effective practices you can implement is harnessing the circular economy.

The Circular Economy: Our Solution

In this book, we will guide you through an actionable exploration of the circular economic model: what it is, how it is (and can be) applied in the real world, and why it matters so much—both for the sake of your business's long-term resilience and profitability and that of the world we call home.

The **circular economy** is an every-market model of production and consumption that facilitates the sharing, leasing, reusing, reconditioning, repairing, refurbishing, and recycling of existing materials and products for as long as possible.

Sounds simple enough, right? But, in our experience, shifting to participation in a circular economic model is seen by many business leaders as a major hurdle. We blame this on the fact that the traditional linear economic model is so firmly ingrained. (That is, the *take, make, use, destroy* cycle we all grew up with.) It's systemic. That said, the opportunity cost here is simply too great to ignore. Bring us a fellow business leader who *isn't* concerned about rising equipment costs, inventory scarcity, lack of confidence in their supply chain, constantly evolving corporate standards, finding and retaining talent, and cash flow, and we'll have to ask if they're an ostrich—because their head must be stuck in the sand.

Let's face it: in a landscape where innovation, regulation, and consumer preferences are ever-changing, stasis is the path to obsolescence. As much as we can empathize with the "If it ain't broke, don't fix it" mantra as business owners, we're waving the flag: *It is broken.* The linear economic model will lead you straight off a cliff if you follow it to its natural end.

We're sounding the alarm, yes, but just like you, we don't like to go on about problems without talking about solutions. And we have a new path to show you. You don't have to be part of the unsustainable status quo, participating in practices that aren't good for you or the environment, any longer. A greener, better, and even easier solution exists—and it's not nearly as big of a hurdle getting from here to there as you might think.

Where You Come In

No matter whether you work in biotech or food manufacturing, construction or retail, in a hospital or a public school, you can bring the principles—and benefits—of the circular economy into your workplace. We've built a thriving enterprise around the concept and are confident that there's a place for it in every industry.

Most people we talk to have a vague idea of what sustainability is about. Maybe they participate in the occasional green initiative around the office. But for many, there is so much largescale opportunity still lying ahead. Even if it's not and never will be your entire business model, sustainability can be something you drive in your organization, from any position within it— or it could even be a role that you carve out for yourself! Imagine your new business cards emblazoned with the title Director of Sustainability, or even Chief Sustainability Officer. (If you own your own business, you're of course welcome to add that title to your business card, too.) If any of what we've talked about has sparked your interest, it's time to get started.

Our journey begins with a deep dive into the circular economy in chapter 1, noting key differences between it and the traditional linear economy, as well as explaining the many opportunities that are out there for putting the circular economy into practice. Chapter 2 explores innovations that are driving the circular economy forward, and the trends and strategies you can use in your own organization. Chapter 3 focuses specifically on the benefits of equipment recovery. Then, chapter 4 looks at the potential of the circular economy and circulatory asset management within and across a range of industries (with real-world examples to show how much potential there is and what companies are already achieving). In chapter 5, we bring up another pivotal piece of this work: the way these initiatives translate into *your* business's sustainability agenda.

Next, we all know how any mention of AI can result in feelings of trepidation, but when talking about integrating AI into your circular economic practices, it can be a simple tool for saving your organization time and money. We'll show you how in chapter 6. Speaking of systems and tools in need of standardization, international standards in the circular economy are lagging, just like those related to AI; chapter 7 calls on the greater community, including manufacturers and governments, to get up to speed. We'll look at what certain governments are doing thus far and what we'd like to see next, based on our extensive experience in the space. Finally, chapter 8 will help you become part of something bigger by empowering you to spread this newfound knowledge to other members of your team, organization and broader network, ultimately bringing the power of the circular economy to the next level.

By the end of this book, you will be confident that you understand all areas of the circular economy—and you will be fully equipped with the tools and strategies needed to implement it in your organization and inspire your teams throughout the transition. Our shared goal is twofold: that you will be able to save money for your business in the short and long terms, and that professionally, you will be confident you're doing what it takes to save our world that is in desperate need of change!

Let's get started.
— Justin Andrews

Chapter 1

The Circular Economy: What's the Big Deal?

O ur first multimillion-dollar project as business owners involved managing asset recovery and site shutdown for a client. A big part of the job was removing high-value R&D equipment, like DNA sequencers and robotic liquid handlers. Another was handling the uninstalling of facilities equipment, like huge chillers and a backup generator, as well as details like removing HEPA filters in the ceiling before demolishing a handful of cleanrooms. The client had hired multiple subcontractors for various pieces of the shutdown workload. Over the few months that it took to complete the project, there were times when there were more than 50 workers on site.

From that work, we recovered several large truckloads collectively worth millions of dollars in equipment, most of which would typically be scrapped during demolition. But we understood this equipment not only represented millions of dollars in revenue for us, our partners, and clients on the resale market—because we kept these machines in use—but also thousands of pounds of landfill avoidance and (literal) tons of carbon avoidance. Environmentally speaking, because we weren't dumping the equipment, disposal-related fossil fuels weren't spent. Plus, the companies that had bought the used machines did not need to buy anything new, so no raw materials were required for them to continue or expand their own operations.

The project, as with all the projects we undertake at REUZEit, ended up producing more cash from the sale of equipment than all the costs required for our team to manage the project. That's the beauty of circularity: by connecting equipment at the end of its first life to the start of a new life, you create both financial *and* environmental sustainability. It's truly a no-brainer business decision for any company.

What Is the Circular Economy *Exactly*?

As we covered in the introduction, the circular economy is a reimagined model of production and consumption that involves sharing, leasing, reusing, reconditioning, repairing, refurbishing, and recycling existing materials and products for as long as possible.

Another definition: According to Ron Gonen in his book *The Waste-Free World: How the Circular Economy Will Take Less, Make More, and Save the Planet,* "It's an economy that invests in advanced technologies related to material science, product design, recycling, and manufacturing that leads to a zero-waste 'closed-loop' system in which resources are not wasted."[8]

The model has three key principles, encompassing the entirety of a product's lifecycle, as laid out by the Ellen MacArthur Foundation[9]: 1) **eliminate** waste and pollution, 2) **circulate** products and materials, and 3) **regenerate** nature. Let's dig into each below.

Eliminating waste and pollution—how we make products

The first step in breaking free from the take-make-waste system of the linear economy—i.e., you take resources, you make a new product, and then you throw the product away when you're done with it—is to *design* products in a way that eliminates waste and pollution. Rather than designing for disposability (meaning the raw materials used to make something eventually end up in an incinerator or landfill, their usefulness

31

forever lost), the circular economy mindset asks that manufacturers and product designers start with materials that can be reused—that can re-enter the economy. Put simply, the goal is to strategize for circularity from the very beginning.

The single-use packaging you see everywhere is a prime example of a product model ripe for new thinking. Luckily, many great minds have already begun addressing this problem. From a mushroom-based, compostable replacement for Styrofoam that can be used to protect delicate products in shipment to edible food packaging made from seaweed, creative and effective solutions exist. They simply need wider marketplace adoption.

Eliminating waste doesn't just prevent a loss of raw materials—it also plays a huge role in eliminating pollution. The Ellen MacArthur Foundation's climate change report[10] found that applying circular economy strategies to product design models in five key industries—cement, aluminum, steel, plastics, and food—would eliminate more than 20% of emissions from the production of goods, equivalent to 9.3 billion tons of carbon dioxide by 2050. Thus, the two key concepts work hand-in-hand.

Circulating products and materials—how we use products

The second principle of the circular economy specifies that we use things in a way that *preserves their highest value*, rather than using them up. The goal is to keep goods in use for as long as possible—as a product first, and as raw components or raw materials next— thereby conserving energy, labor, and materials.

Consider a toaster. The idea here is to design a toaster that can be a toaster for as long as possible, for a toaster is *more valuable* as a toaster than it is as disparate component parts. In this instance, circularity begins with building a durable appliance that can be easily repaired, and perhaps creating a network that customers can use for maintenance, repair, and refurbishment. Modular design would allow for replacement of individual parts as opposed to replacing the entire toaster when one thing breaks. And when it comes to circulation of products, a toaster manufacturer that's thinking through the lens of the circular economy would make it easy for a customer to return a faulty, broken, or no-longer-wanted toaster for subsequent reuse.

Toasters are part of what's known as the "technical cycle." The other group of materials that make up our economy is referred to as the "biological cycle;" these are materials that can safely re-enter the biosphere. Food byproducts, for example, can be returned to nature by composting them. (As a way to regenerate more food, composting is a circular act.) Take wooden furniture as another example. Wood is biodegradable, but screws are not. Paint and glue can biodegrade, depending on the furniture design team's choice of materials. (Thinking back to the previous category of eliminating waste from the point of design, circularity requires intentionality on behalf of the product designer(s) so that these components can be broken down easily and will re-enter the economy at their highest use.)

Another way to keep products in use—and to maximize their usefulness—is through the sharing economy. Examples of this include furniture or equipment rentals, car sharing services, and special-event clothing rental websites. We'll explore this in depth in the chapters to come.

Regenerating nature

Nature regenerates itself. But we've inserted ourselves into natural systems in order to do things like grow food. In other words, waste is a human invention. The goal of this third principle of circularity—regenerating nature—is to ensure that when we interact with nature as human beings, we work to regenerate whatever we deplete. In fact, the focus should be on actively improving the environment rather than simply reducing harm. Consider regenerative farming practices. Their intention is to improve soil health and design agricultural environments that approximate natural ecosystems. Furthermore, designing and building a regenerative environment in urban areas—by incorporating things like green roofs—both makes buildings more efficient and brings elements of a natural habitat into cities.

Circular Economy Systems

The Ellen MacArthur Foundation illustrates the circular economy ecosystem with their butterfly diagram, below:

RENEWABLES FLOW MANAGEMENT

RENEWABLES

FINITE MATERIALS

STOCK MANAGEMENT

REGENERATION

BIOSPHERE

BIOGAS

ANAEROBIC DIGESTION

EXTRACTION OF BIOCHEMICAL FEEDSTOCK[2]

BIOCHEMICAL FEEDSTOCK

FARMING/COLLECTION[1]

CASCADES

COLLECTION

CONSUMER

COLLECTION

USER

SERVICE PROVIDER

PRODUCT MANUFACTURER

PARTS MANUFACTURER

SHARE

MAINTAIN/PROLONG

REUSE/REDISTRIBUTE

REFURBISH/ REMANUFACTURE

RECYCLE

MINIMISE SYSTEMATIC LEAKAGE AND NEGATIVE EXTERNALITIES

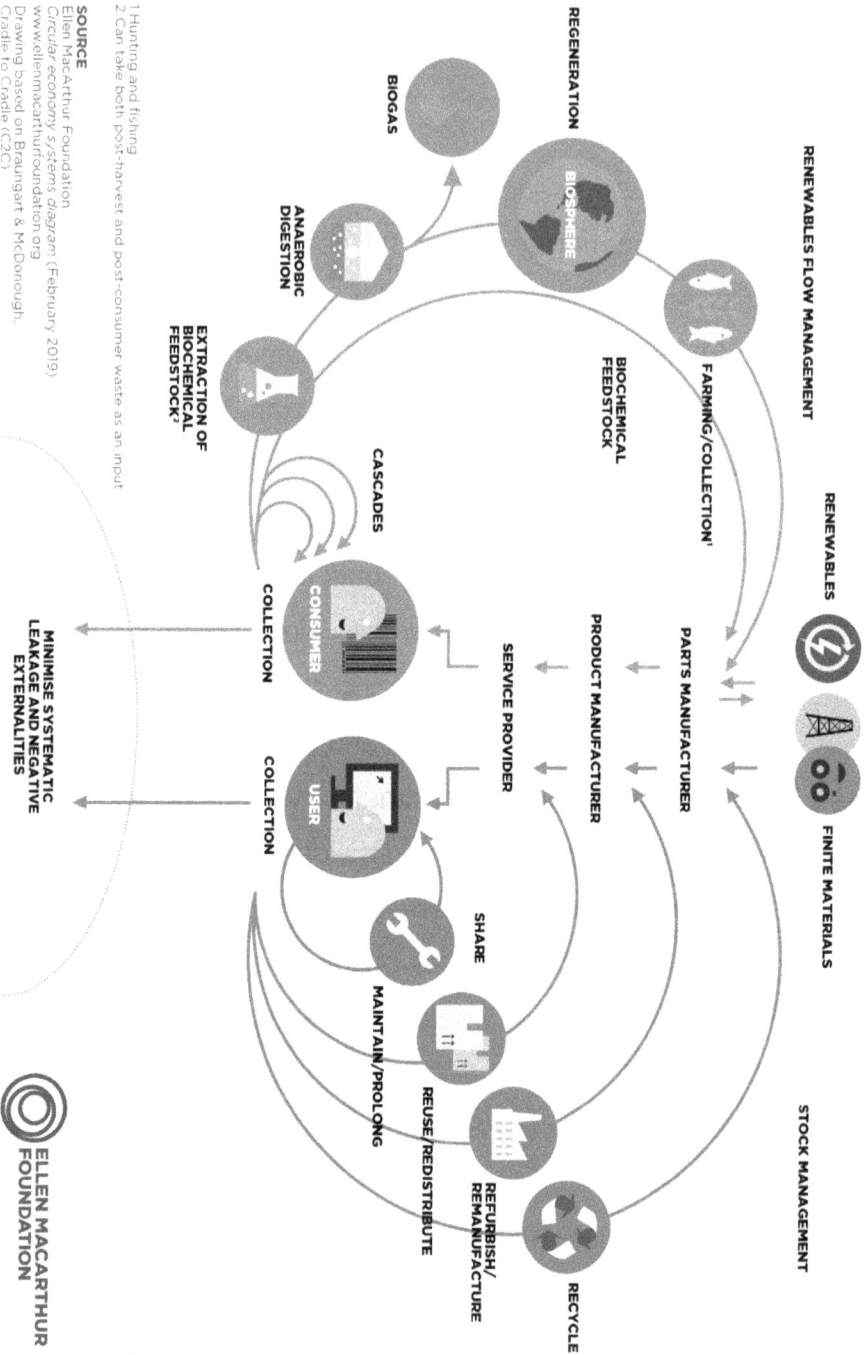

1 Hunting and fishing
2 Can take both post-harvest and post-consumer waste as an input

SOURCE
Ellen MacArthur Foundation
Circular economy systems diagram (February 2019)
www.ellenmacarthurfoundation.org
Drawing based on Braungart & McDonough,
Cradle to Cradle (C2C)

ELLEN MACARTHUR FOUNDATION

Figure 1.1

36

The underpinnings of these material cycles—referred to as renewables and finite materials above, and in other places as the biological and technical cycles (as previously mentioned)—include the following key tenants of circularity, moving (mostly) from highest value to lowest value:

- **Maintain/prolong (and share):** Within the technical cycle, this concept refers to keeping materials in use for as long as possible by designing for durability and easy maintenance and repair. Products with a longer lifespan also have greater potential to be shared.
- **Reuse/redistribute:** The idea here is similar to a secondhand market—keeping products in use without changing them, or with minor repairs.
- **Refurbish/remanufacture:** Refurbishing a product keeps it in use by restoring its value—repairing it without taking it apart or replacing parts. Remanufacturing involves disassembling a product and rebuilding it, including replacing components as needed. A remanufactured product is put back into the marketplace in as-new condition.

- **Recycle:** Recycling breaks a product down to its material level with the goal of using as much of the materials as possible in the production of new products. *Recycling is important, but it is often overemphasized in conversations around sustainability.* As it requires labor, energy, cost—and usually involves some materials loss—it is therefore a lower-value process than those described above.

- **Cascades:** Within the renewable or biological cycle, a cascade entails using materials and components again and again, usually for increasingly lower-value applications, until they ultimately need to be returned to the environment. Continuing with the wood furniture example from above, think back to the source material: wood from a tree. The wooden furniture is made from the highest quality wood. Lower quality wood could be used for construction framing, and scraps could be used as particleboard, wood chips, or burned for fuel.

Most of our work at REUZEit falls within either the reuse/redistribute cycle, when we make minor repairs or small enhancements to equipment, or the refurbish/re-manufacture cycle, when a machine needs some larger cosmetic or functional updates before re-entering the marketplace.

This all goes back to our earliest days as a company. One of our first projects was completely overwhelming on paper: our huge client had just merged with another large company, and they were planning to build out a new 1,000,000-square-foot building. The space they'd acquired for this purpose had been previously used as a storage facility—or, more accurately, an equipment graveyard. Our job was to clear out a 200,000-square-foot warehouse that was full of equipment. The client gave us the keys and told us everything had to be gone in three months.

We got to work immediately, organizing and processing each item. We created a temporary processing area to test and recondition equipment, take photos of the updated machinery, research each item to establish an appropriate resale price, and list everything for sale on eBay. Once we sold something, we packed and shipped it in this same space.

Working long days, weekends, and nights, we managed to sell *almost* all the equipment in our allotted three months. The handful of things that we couldn't sell in time were recycled, and a few items were brought to our new "warehouse"—which was simply a 1,000-square-foot flower shed in the agricultural district of San Marcos, California. (Justin had found this space because he'd spent his undergrad years at California State University San Marcos, where he'd learned that farmers rented out their warehouses.)

It was our first foray into proving that the benefits of the circular economy could beat out the status quo linear economy—which would have seen the client simply calling in a team to dump all that equipment, writing a fat check to get it out of sight, out of mind. But in working with us, they actually *made money*, while putting equipment that had been gathering dust back into use and making space for their new business venture at the same time.

Circular Economy vs. Linear Economy

We've been referencing the linear economy regularly thus far. Here, we'd like to take some time exploring how exactly the circular economy differentiates itself from the linear status quo.

A linear economy is based on the "take-make-waste" model, i.e., a system in which a business takes raw materials and makes something from them, which citizens then consume and throw away. The term might be new to you, but it's a concept you're all too familiar with. Think of single-use coffee cups, paper plates, and/or so much of the packaging you interact with daily: the netting that bunches clementines into easy groups at the grocery store, plastic bags filled with shredded cheese, Styrofoam to-go boxes or packaging fillers. And on and on.

But let's get specific. Overall, the two economic models differ in terms of their respective approaches, visions, takes on sustainability, and perspectives on how businesses operate:

- **Approach:** The linear economy uses the take-make-waste approach. The circular economy adheres to the 4R principles of "reduce, reuse, recycle, and recover," a closed loop approach that feeds itself, in which there is no form of waste.

- **Vision:** Linearity aims to drive short-term revenue growth, via whatever means necessary (planned obsolescence, drumming up demand, mass production, et cetera). The circular approach adopts a long-term view, one that makes sustainability and extending product lifecycles possible while also maintaining or even furthering long-term *profitability* (as opposed to measuring revenue, an important distinction).

- **Sustainability:** The linear status quo's stance on sustainability is to "reduce harm." Circularity doesn't merely aim to minimize the negative impact of production on the environment; eliminating waste from the product lifecycle is just the baseline. Rather, the circular model works to *maximize* environmental benefits and resource value.

- **Business models:** Traditional manufacturers working within the linear mindset focus on making products they will never touch again, after those products leave their facilities. (Consumers will take them from there, use them, and eventually throw them away—a pattern of no concern to the original manufacturer, an entity that has made itself both distant and distinct from its products' downstream eventualities.) While product manufacturing absolutely exists in the circular economic model, it follows a service-oriented model that involves thinking through the experience of using the product *during the design process*. That is, manufacturers see themselves as having a broader role within the product lifecycle, purposely and strategically designing for product durability, easy maintenance and repair, sharing, reuse, or upcycling down the line.

In short, whereas the linear model focuses on profitability—with zero regard for a product's lifecycle—the circular economy focuses on sustainability. And this is a crucial point: circularity still allows for production. It just strategically prioritizes and designs itself around a model of production that generates the least amount of environmental impact.

The 14 Tangible Benefits of a Circular Economy

The circular economy isn't just nice in theory. It offers real, tangible benefits that are ready for the taking. Citing real industry examples, we've highlighted 14 of the strongest arguments for switching to this model below—for both business *and* environmental impact.

Business benefits

1. Develop new businesses or product lines from waste

Yarn producer Aquafil Group discovered that Nylon 6—a material used in fishing nets, fabric scraps, carpet flooring, and industrial plastic, and that often ends up in landfills—can be recovered and converted into new yarn with no loss of quality[11]. They call this product ECONYL® regenerated nylon. To date, it has been used in everything from chairs to sports bras, tote bags to rugs. Aquafil's biggest challenge has been in reliably reclaiming the material, to which end they've built a qualified program with strict requirements on suppliers' environment protocols. Overall, a material that had previously been thought of as waste has become a new raw material to feed new product lines in their production process.

2. Create new opportunities

Economic growth is not dependent on resource consumption. For those of us who've grown up under the guise of the linear model, this bears repeating: *we don't have to keep making more and more stuff in order for the economy to thrive.*

Put another way, many new and even superior opportunities for businesses exist within the circular economy. Specifically, in the circular model, growth is fueled via regeneration, increasing an ecosystem's yield over time and capturing more value from the infrastructure and products we already have. A 2015 McKinsey report found[12] that adopting a circular economic approach would allow Europe "to grow resource productivity by up to 3 percent annually" and generate, via combined primary resource and non-resource externality benefits, around $1.8 trillion per year for Europe's economies by 2030 (versus the time of the study).

One concrete example is Novaxia[13], a French investment company that works to convert vacant commercial buildings—the number of which have greatly increased since the COVID-19 pandemic—to low-carbon residential properties. This "urban recycling" mission provides housing while reducing urban sprawl.

3. Drive innovation

Innovation, by definition, is a new idea, method, product, device, et cetera. It can also be thought of as the introduction of something new.[14] Therefore, it's not surprising that when people and organizations begin to shift their mindset toward an economic model that is new to them—in this case, the circular economy—innovation often follows.

Consider circular innovation in the industry of building design and construction. While circularity focuses strongly on reuse, there's always going to be a need for new things. The strategy around the creation of these new things should be a focus on optimization, and this industry is primed for circular innovations that will help close the loop of linear consumption. Here are some examples of how this industry can optimize:

- Designing for modularity to give buildings the potential for longer life.

- Creating low-carbon structures to minimize their long-term environmental impact.
- Using prefabricated and upcycled materials, as well as new technologies like 3D printing, to decrease reliance on raw material use.
- Working with robotics to improve lead times and reduce dependence on fossil fuels during the building process.

Companies across the world—including Hyperion Robotics in Finland, Gropyus and CREE GMBH in Austria, BoKlok in Sweden, and Daiwa Modular House in Japan— are leading these efforts.

4. Create more jobs

As deputy commissioner for sanitation, recycling, and sustainability in New York City during the Bloomberg administration, Ron Gonen and his agency launched a curbside food-waste collection program. They recaptured food waste that had previously gone to landfills, either converting it to rich compost that was sold to landscapers or turning it into natural gas via anaerobic digestion. The initiative not only eliminated the city's $150 million annual expense of sending food waste to landfills in other states; it also created thousands of local jobs in the composting and anaerobic digestor industries. [15]

5. Make businesses more resilient

When a business is no longer dependent on an unreliable supply chain, it becomes more resilient. Simple as that. Why? Factors like fluctuating raw material prices or extreme weather events disrupting transportation have less of an impact—or no impact at all—on organizations built to accommodate reused and recycled materials or products.

6. Provide consumer benefits

Many consumers today look to support companies that value the same things they do. In an increasingly volatile world, it simply feels good. Refreshing, even. In the technical product cycle, durability and easy repair/maintenance allows people to keep the things they buy (or rent, or lease) longer, increasing their disposable income in the long run. This makes a customer feel like a company understands what and how things matter to them. When we consider the biological product cycle, using regenerative farming practices, for example, is good for the farmer's soil and land *and* pro-vides consumers with healthy foods free from things like harmful chemical pesticides. A practice like this in-dicates to the customer that their health matters to the people producing their food. Everyone benefits.

7. Increase demand for services

There are two ways to think about how circularity in-creases demand for services (as opposed to products). First, in essence, as more organizations move to the circular economy approach, more CE-specific services will be needed to keep products and materials moving through the loop. This includes companies or individu-als focused on maintenance, repair, refurbishing, and remanufacturing, all highly skilled roles; platforms to sell, rent, or share reused products; and organizations that will manage reverse logistical efforts to recapture materials and/or reintroduce products into the system. Second, circularity asks that we reimagine the value of a product—not merely thinking of it as a thing, but as a thing with a *use*. Swiss architect, founder and director of the Product-Life Institute Geneva, and sustainability pioneer Walter Stahel champions selling a product's performance, referred to as product-as-a-service. In the purest form, this means that the manufacturer essentially rents out a product, profiting from its use only—thereby incentivizing said manufacturer to design for longevity and easy repair.

In both cases, the key takeaway that separates this benefit from others in this list is the human resources impact. Service-oriented jobs like those required by the circular economy often require a strong skillset acquired via experience and/or training. This reorientation towards prioritizing service within our economic landscape will result in the creation of more highly skilled jobs—and more companies to employ these skilled professionals.

8. Foster collaboration

One of the circular economy's biggest challenges is also one of its greatest rewards: in requiring collaboration, it fosters a sense of mutually beneficial interdependence. Everyone has a role to play in this system, including entrepreneurs and business owners, product designers and innovators, policymakers and governments, investors and financial institutions, suppliers and retailers, and, of course, end consumers.

Responsibilities being spread out also means more people have the chance to be a part of the model's success.

Furthermore, the business model requires a continuous dialogue throughout a product's lifecycle, increasing customer intimacy. There is no downstream, no "throwing *away,*" no "out of sight, out of mind." That is, manufacturers don't just pass something on to the next step in the process, leaving any follow up to downstream partners; rather, they remain engaged. A major benefit of this is built-in consumer feedback, about things like consumers' usage patterns with a particular product. This data can then inform future iterations of the product, improving its functionality and longevity— and, potentially, profitability.

Environmental benefits

9. Carbon Avoidance
Regardless of CO_2's potential influence on climate, the carbon footprint of equipment production represents the total pollution involved in raw material extraction, processing, manufacturing, logistics, and use. These impacts can be significantly reduced by applying

circular economic principles, streamlining operations and measuring carbon reduction as part of a net-positive financial strategy. That's been my professional experience at REUZEit. The total carbon footprint of new equipment entering and re-entering international markets is immeasurable, but the more we reuse equipment, the less demand there is for new production. The environmental benefit of this approach is substantial and offers a more direct, scalable reduction in pollution than many current sustainability initiatives, including carbon-free electricity, eco-buildings, or automated transportation technologies.

Here's the exciting news: circularity specifically could make a huge dent. According to the 2021 Circularity Gap Report[16], switching to a circular economy could reduce carbon emissions by 39% in just over 10 years.

10. Increase recycling and reuse

This one is obvious, but it merits direct discussion. The circular economy is largely focused on increasing rates of reuse and recycling, and the environmental impact of such efforts is huge. According to the United Nations Environment Programme's Global Waste Management

Outlook 2024 report[17], if the world were to adopt a circular economy approach, worldwide municipal solid waste could reduce from more than 4.5 billion tons to less than 2 billion tons a year by 2050. *And* there would also be a hard cost savings associated with this impact: while the projected *cost* in 2050 for continuing on our current solid waste management path would be $640.3 billion, transitioning to a circular economic model would amount to a *net gain* of $108 billion. And this is just in the realm of solid waste management, which is only one piece of the puzzle.

Contributing to this scale of impact, circular economy approaches aren't relegated to small startups. Household brand names are making strides, too, proving that the model is gaining mainstream momentum. In 2020, IKEA opened their pilot secondhand store in Sweden, featuring gently used furniture that had been repaired in an adjacent shop. And in 2024, they launched their online "as-is" platform (initially branded as IKEA Pre-owned with test markets in Madrid and Oslo), expanding the "second chance" market to anyone who lives close enough to a store that's offering the program and can pick up pieces in person. IKEA is targeting 100% circularity by 2030[18], and other brands are following suit. In 2023 in the US, the popular kids' clothing brand Hanna Andersson launched its own resale site, fondly known as Hanna-Me-Downs[19].

11. Reduce the impact of manufacturing

The manufacturing industry's negative impact can be traced back to the times of the Industrial Revolution. In 2002's landmark text *Cradle to Cradle,* authors William McDonough and Michael Braungart discuss how a new city life proliferated under industrialization, from the demand for "more, more, more—jobs, people, products, factories, businesses, markets..." The air in Victorian London "was so grimy from airborne pollutants, especially emissions from burning coal, that people would change their cuffs and collars at the end of the day (behavior that would be repeated in Chattanooga during the 1960s, and even today in Beijing or Manila)."[20] Today, much has been written about how those who are most impacted by pollutants—(see Elizabeth Rush's powerful and heartbreaking *Rising: Dispatches from the New American Shore)*—are those with the least amount of privilege. No one anywhere in the world should have to deal with pollutants in their air or groundwater due to neighboring manufacturing

54

facilities—but that doesn't mean such facilities have to go away. In fact, reimagining these spaces as functional parts of their local ecosystems, rather than drivers of displacement, is part of circularity.

One of the concepts that we're inspired by is that of industrial symbiosis. In the Dutch city of Rotterdam, BlueCity is an industrial symbiosis hub. Developed from a vacant swimming pool in 2015, BlueCity is now home to 55 entrepreneurs focused on circular businesses and innovations. (It was created by the founders of Rotterzwam, which collects coffee grounds from area businesses and uses them to cultivate mushrooms. Those mushrooms generate carbon dioxide, which Spireaux, another BlueCity company, uses to make an algae-based paste for veggie burgers.) Other startups in the hub are using the city's waterways to inspire efforts toward a blue economy, the "sustainable use of ocean resources to benefit economies, livelihoods and ocean ecosystem health"[21] (hence the name of the hub).

12. Reduce energy costs

ING found, in a 2019 study, that 59% of organizations that were trying to learn about the CE framework and 49% of those that planned to move toward a CE model were motivated by achieving cost savings, as either their first or second most important motivator.[22] Further, improving resource efficiency and reducing waste were the leading factors driving companies' sustainability practices. A primary way that circularity helps save costs and improve efficiency is by reducing energy costs.

Specifically, a CE business model reduces energy costs by reducing energy consumption in the manufacturing process, carbon gas emissions, transport and extraction costs, and reliance on virgin materials. A 2017 study[23] that looked at the implementation of a specific suite of circular economy approaches found that they would result in a 6–11% reduction in energy use required for economic activity worldwide. Think about how that type of baseline cost savings could impact your P&L!

13. Mitigate land degradation

The health, livelihoods, and security of an estimated 3.2 billion people worldwide are at risk due to the impacts of land degradation[24]. Circularity aims to mitigate this impact by increasing land productivity and enhancing soil quality by returning nutrients to the soil as part of the closed production loop.

One example of a company working to assuage land degradation through the creation of a circular system is the American personal care company Dr. Bronner's, and their subsidiary Serendipalm[25]. Dr. Bronner's uses palm oil as a main ingredient in many of their personal care products. Palm oil sourcing has become notorious for its environmental destruction and harmful impacts on local communities—and that's why Dr. Bronner's started their own ethical and nature-positive palm oil sourcing project, Serendipalm, in 2007.

Serendipalm works with smallholder oil palm farmers in Ghana, training them in methods of regenerative organic cultivation and dynamic agroforestry—i.e., integrating crop production within forest systems, in this instance by growing secondary crops (like bananas, turmeric, and ginger) alongside palm and cocoa. And additional circular efforts have been implemented throughout the Serendipalm business cycle. They only use the outer palm fruit, so the palm kernels are collected and sold to a Togo-based fair-trade soap production company in a neighboring town. Crop byproducts are mulched into fertilizer and cycled back to the farms, returning nutrients to the soil. Finally, Serendipalm has helped develop and disseminate locally adapted, high-yielding palm tree seedlings, which are more resistant to pests and better suited to local growing conditions.

14. Promote natural resources conservation

As Ron Gonen so eloquently summarized in *Waste Free World* and British economist E. F. Schumaker argued in his 1973 book *Small Is Beautiful*, "Nature's bounty should not be thought of simply as resources…but as the fundamental capital that makes all of our production and consumption possible."[26] In Schumaker's words:

> A businessman would not consider a firm to have solved its problems of production and to have achieved viability if he saw that it was rapidly consuming its capital. How, then, could we overlook this vital fact when it comes to that very

big firm, the economy of Spaceship Earth and, in particular, the economies of its rich passengers?[27]

The linear economy necessitates the constant extraction of finite resources, with no end in sight. Extending material lifecycles in the circular economy will cut down on this unsustainable extraction process, even as our international population continues to grow, and, in turn, will reduce the energy consumption associated with never-ending production. And reducing our consumption patterns ties directly to conservation. As such, many organizations are taking things a step further by not just conserving (i.e., reducing harm) but *regenerating* (i.e., working toward active improvement) and *restoring* farmland, grassland, and forests—the ultimate objective.

Opportunities for Circular Economy

According to the World Economic Forum[28], the circular economy is the business opportunity of our time. For one, it's an undeniable necessity—we cannot sustain our current consumption levels using the finite resources of our world. In fact, if we stayed on the trajectory that we're now on, by 2050 we'd need 2.3 times more ecological resources than are available in the entire world. Only a shift to the circular economy would decouple our growing population from this pattern of consumption.

As it stands today, we have a long way to go. The Circular Economy Indicators Alliance found in a 2020 report that our world economy is only 8.6% circular.[29] While this statistic might be depressing on the surface, we can choose instead to be inspired by the opportunity it implies: that is, the outstanding *opportunity* to convert 91.4% of the world to circularity. Imagine what an impact that could have! It really makes you want to get in on a piece of the pie.

One giant opportunity exists within the textiles industry. Each year, textile production uses around 4% of the world's freshwater supply.[30] By now, many of us have heard of the detrimental effects of the "fast fashion" phenomenon. But you may not have fully comprehended that the trend led to a doubling in textile production in the first 15 years of this century. Changing this pattern will require a massive mindset shift; yet the potential gains are huge. Efforts such as reducing new clothing purchases, keeping garments in use for longer via secondhand and rental exchanges, and collecting materials that are at the end of their life for recycling or repurposing all add up to a $500 billion economic opportunity in which both individual consumers and clothing companies can partake.

Lastly, moving toward a circular economy will help protect human health and biodiversity. It's estimated that air, water, and soil pollution is responsible for 9 million premature deaths annually[31], making it the largest environmental cause of disease and death in the world today. And pollution, deforestation, and other detrimental methods of resource extraction are leading to a loss of biodiversity that's hundreds, perhaps thousands, of times faster than a natural rate[32] of decline. Scientists have been sounding the alarm for a while now. Fortunately, circular models can make a measurable difference. By making better use of natural resources, reducing energy usage and waste, protecting water and land, and promoting the reuse of products, circularity helps mitigate the pollution crisis and keeps hazardous materials from leaking into the natural environment.

Let's look at the system-wide impact of transitioning to a circular economy for the plastics industry, in particular. The Pew Charitable Trusts found[33] that using existing solutions—including reducing plastic production, using compostable materials in the place of plastic, and improving recycling efforts for the plastics that must remain in use—could reduce the annual flow of plastic waste into the ocean by 80% in 20 years. This is beneficial for human health and biodiversity, of course, but it would also benefit industries like fishing and tourism, as the total natural capital cost of plastic littering to marine ecosystems is $13 billion annually[34]. (Such efforts would lower worldwide health care costs, too, because of reducing pollution and toxic emissions.)

Overall, across industries, research shows that the circular economy could unlock $4.5 *trillion* of economic growth worldwide, because of the cost savings from the associated reduction in waste as well as the opportunities for innovation and employment.

The REUZEit Revolution

A decade ago, we at REUZEit arrived on-site to meet with a client with a giant assignment and a tight timeline. Their task was to shutter an entire campus, which consisted of 470,000 square feet of R&D labs. The employees on the team shared their frustrations with us: they were upset that so much equipment and office furniture, which had cost so much upon initial purchase, was going to waste. They were talking about the monetary expense, of course, but we were also thinking about the wealth of natural resources that had been extracted to produce such equipment, now destined for a landfill—the environmental expense. Prior to discovering services like ours, large companies usually ate the cost of this "dumping" as part of a site closure, because they assumed it was the fastest and most efficient way to handle things. Another example of the status quo causing missed opportunities for recovery, reuse, and direct cost savings.

What did we do instead? We were able to remove *all* the valuable equipment, recapturing it for our reuse and recovery program. To further reduce waste (and save on disposal costs), we donated truckloads of consumables and furniture to local universities and schools. We removed miles of CAT5 wire worth many thousands of dollars and recycled truckloads of scrap

63

metals. All materials that needed to be recycled were recycled locally.

The result? This approach to recovering value and finding new home for "trash" was massively successful. The client had budgeted a $1 million *expense* for this site shutdown. We delivered a *net positive recovery* of $700K. In other words, for this client, we delivered a $1.7 million surprise cost avoidance and recovery. We received a special award as a vendor on this project, for going above and beyond, which we now attribute for securing our relationship and international master service agreement with our top client. That's when our company really started taking off.

The task of adopting circularity impacts governments, consumers, and company owners equally. As leaders in this space, we know firsthand how difficult it can be to go it alone—to take the necessary, concrete actions for such a shift upon yourself—due to the overwhelming amount of change that has to happen: from persuading stakeholders to embrace and champion a shift in operational mindset to guaranteeing strong consumer demand for circular products to the downstream logistics of enacting each and every new strategy. Most people will tell you it helps to break things down into small steps, like cutting waste or applying circular design principles in individual departments. Still, even once someone does grasp the advantages of circularity, they find themselves with a lack of practical platforms to enable streamlined reuse. (Or, as we found with that early client we just mentioned, some employees don't even know about the services that currently exist and that are partnering with their organization. Today, we work on client sites all over the world, and we regularly meet staff who learn about our service and wish they'd known sooner that their organization had a REUZE program.)

Stick with us, because we're going to get into the meat of how to enact change from within your organization in Chapter 8. For now, we want you to keep in mind that there is an active movement toward making the shift to circularity—and an existing platform ready for you to leverage to drive needed change. It's more than any one person, product cycle, or business, small or large. You can get in on the ground floor right now by joining us in the REUZEit Revolution. Learn more at reuzeit.com/reuzeit-revolution.

Get Started: Self-Assessing Your Circular Economy Readiness

We're going to finish out every chapter in this book by giving you tangible steps that will help you Get Started in bringing circularity into your life—whether that's at home or in the workplace.

At this early stage of learning about CE, we don't expect you to start measuring any of your current consumption habits or processes. But one thing you can do is self-assess your CE readiness by using the following tools:

- Circle Assessment: https://cat.ganbatte.world
- Circular Transition Indicators (CTI): https://cti-tool.com/start-and-understand-your-circularity-assessment-step-by-step/

Now that you understand why a circular economy is more than beneficial—it's, in fact, crucial—and you have a handle on how ready you are to embrace CE, it's time to take a closer look at the circularity strategies being employed and applied across the business world.

Chapter 2

Circular Economy Innovation and Business Strategies

The circular economy is already in operation today, barreling forward and gaining steam, powered by innovative minds and organizations across the world. In this chapter, we're going to look at myriad different business trends and strategies you can learn from, apply, and use to take steps toward circularity in your organization. Beyond compelling anecdotal evidence, we also want to let you in on groundbreaking research that's laying a quantifiable foundation for this strategy worldwide.

As business owners, we partnered with thought leaders in the field of closed-loop systems and circular economics to introduce the strategy of CE-SAM (Circular Economic–Surplus Asset Management) as a business approach to creating value by extending the lifecycles of discarded useful equipment. Harold Krikke (who's on the Faculty of Management at Open Universiteit Nederland, and who regularly publishes on these topics[35]) and Nestor Coronado Palma (Managing Director of

Value Loops, a Dutch consultancy focused on sustainable technologies) coauthored the 2022 report on our findings, which was published in the academic journal IEEE Engineering Management Review.[36]

Our goal was to quantify the potential of largescale reuse programs by looking at the real-world equipment we handle in our operations—specifically, by measuring how much carbon gas avoidance these methods make possible. (As we know, carbon emissions are correlated with both overall pollution and use of natural resources.) In sum, we can measure an item's *burden* (pollution, resources, labor) by measuring all the components, materials, processing, and logistical effort required to get that item from raw materials into use.

We distinguished four types of value promoted by CE strategies and methods: 1) sourcing value, 2) customer value, 3) informational value, and 4) environmental value. *Sourcing value* refers to the direct financial gains that come from sourcing less expensive resources, avoiding the use of new resources, and avoiding disposal fees and environmental fines. The cost to source remanufactured products, for instance, is about 40% lower than for newly built products. Next, *customer value* indicates activities that enhance customer satisfaction and loyalty via well-organized channels that allow for simple returns at the end of a product's life. Third, *informational value* comes from a business's ability to analyze returns data on common production or supply disruptions, product failures, a product's useful lifetime, consumer usage patterns, and complaints. This information is extremely valuable for improving business processes. (We touched on this last chapter, when we discussed how fostering collaboration is a major CE business benefit.) Finally, footprint reduction, proactive compliance, and environmental leadership through reuse and recycling contribute to *environmental value*.

For this study, we took a deep look at two different items that we frequently refurbish and sell, and that are representative of the two general types of equipment we handle: ultra-low freezers and mass spectrometers.

A ULT (or -80) freezer stores material from below -40°C to as low as -150°C. The pharmaceutical industry uses ULTs to preserve drug compounds and biological samples. A ULT freezer is big but not heavy for its size, because it's mostly open space inside. Because it's not actually made of a lot of material, it has a lower total carbon emission than you might think. We found that a used, late model (less than 5 years old) freezer can be redeployed with or without reconditioning while still maintaining manufacturer specifications at an average savings of $10,000 per product. Approximately 31,320 newly manufactured freezers are expected to be pur-chased in 2027 worldwide. If 50% of these newly man-ufactured products are replaced by redeployed products—a percentage that is in line with European Union targets—buyers and companies will realize di-rect annual economic benefits of more than $16 million, for this one type of product in these industries alone.

Taking things from the hypothetical to the actual, in just one recent month (December 2024), we redeployed 300 ULTs from the UK to the US and the EU, saving about $3M and 62,550 CO2 kg equivalent, or roughly 15 gasoline-powered passenger vehicles (or 55 elec-tric-powered passenger vehicles) driven for one year.[37]

On the other hand, a mass spectrometer is dense and high-tech, often used in concert with other specialized instruments, and made from exotic metals and plastics. It therefore has a higher carbon emission. Medical labs employ mass spectrometry to diagnose metabolism deficiencies, determine whether biomarkers or enzymes are present, and in toxicology testing. Similar to the ULT freezer example, an average used, late-model spectrometer can also be redeployed to meet manufacturer specifications, but in this instance, with an average savings of $150,000 versus buying new. The total market for these spectrometers in 2030 is estimated to be around 31,750 (again, close to the demand for freezers), which equates to a $2.4 billion annual savings if just half of new mass spectrometer purchases are replaced by redeployed alternatives.

In both instances, we calculated recovery options for closing the loop as significantly contributing to a reduction in carbon emissions. A ballpark estimate is that annual reductions of 1.3 million $kgCO_2$ (recycling-based) up to 3.2 million CO_2 kg equivalent (direct redeployment) are realistic for ULT freezers by 2027. Correspondingly, for mass spectrometers, reductions would equal 285,000 $kgCO_2$ eq (recycling-based) up to 1.2 million $kgCO_2$ eq (direct redeployment) annually by 2030. At a high level, in addition to the impressive financial savings outlined above, carbon footprints can be reduced by up to 80% by redeploying equipment, depending on residual value and customer needs.

As this research demonstrates, CE-SAM can be a worldwide economic and environmental game changer, helping organizations reach ambitious sustainability goals via natural and leverageable productivity methods—*if it is integrated as part of a business strategy.* But to be sure, innovations in the circular economy go far beyond our work at REUZEit to redeploy scientific equipment. Across industries, the innovations driving the circular economy start at the source.

Upstream Innovation

Rather than waiting until they have a pile of waste to deal with to implement the principles of the circular economy, many bright minds are using upstream innovation[38] to proactively trace waste back to its source with the goal of preventing such waste in the first place. Tackling the problem at its root cause, as it were. From a product development perspective, this involves rethinking the product, packaging, and business model during the design phase.

Innovation at the product design level can include considerations of formulation, concept, shape, and size that will satisfy consumers' needs and expectations without creating waste. Packaging innovations like format and material choice can also design out waste. (Switching to recyclable or biodegradable packaging is a straightforward example.) Thinking at the *system* design level can mean modifying business models, supply chain networks, and production processes. Prioritizing local sourcing and production is one way to eliminate the need for packaging, as well as transportation dependencies.

Danish footwear startup VAER is one example of an upstream innovator. They reimagined sneakers at the product design level, using textile waste from jeans and workwear, as well as materials from shoes returned by their customers, to produce vegan, sustainable, and truly good-looking shoes. This upcycling process is known as "waste-to-resource" and is considered one of the top up-and-coming circular economy trends[39].

Entire business models can also be, and are, built around the concept of reuse. Excess Materials Exchange (EME) is one example. EME designed a digital matching platform that connects excess/unused materials or waste products to their highest-value next use. Each material or product that's added to the platform is assigned a digital identity that assesses its composition, toxicity, and more. Then, the platform identifies reuse options based on financial, environmental, and social value criteria. EME's goal is to make it easy and efficient for companies to "identify, exchange, and repurpose excess materials, transforming potential waste into valuable resources." They've found their tool increases the financial value of material flows by 110% on average and reduces a company's ecological footprint by 60% on average. In other words, it's effective for both profits and the environment.

Another example of a company with reuse at its core is Vytal, who've dubbed themselves "the reuse system." With a mission and model tied entirely to packaging—specifically, reusable food packaging—they make bowls, trays, mugs, and more from recyclable polypropylene. Their mobile app gives consumers the ability to locate restaurants that allow for free borrowing or returning of this packaging, and all items carry a QR code that partner shops can scan to ensure the packaging is, in fact, returnable. Vytal collects and cleans each piece before putting it back into use.

The internet of things (IoT) is exciting in that it offers tons of opportunity in the world of waste management. In communities across the world, because standard waste collection is usually run on a set schedule, trucks may visit sites to find bins either nearly empty or over-flowing. Greek startup Recytrust developed an IoT-based digital weight scale for recycling bins in an effort to reduce such inefficiencies. The scale is placed under a bin, generating alerts when its weight reaches a certain threshold, indicating that the bin needs to be emptied. (An added benefit for drivers is real-time metrics on the weight of recyclables yet to be picked up.) The technology also gives users the opportunity to monitor their own recycling efforts.

In India, Ishitva Robotics Systems also designed an innovative business model around an IoT-enabled smart bin that automatically segregates dry waste, for ease of sorting during the recycling process. This bin collects data on usage patterns, types and amounts of waste collected, and time spent en route, to inform things like route optimization. UK startup Recycleye takes a similar approach, powered by both robotics and artificial intelligence (AI). Their algorithms replicate human vision to identify, classify, and sort items in waste streams. In the past, the cost of sorting resources outweighed the value gained from selling them, but this automation has increased the profitability of materials recovery by improving efficiency and accuracy. Recycleye has also created a visual database, called *WasteNet,* of waste items labeled at the brand level. They've used resulting data for machine learning purposes, leading to continuous improvement in the robots' sorting capabilities as well as the development of a waste taxonomy that's considered the standard for waste classification.

Building on these industry examples, we'd now like to look at a handful of frameworks that could help you think about how to achieve a future-based circular economic model for your own company. These strategies are all ways of driving upstream innovation, in turn. There's some overlap between and among them, but we've called on all of them at different times in our work. Review them below and decide which makes the most sense for you and/or your organization.

Framework #1: The 10 Rs Strategy

Up first is the **10Rs strategy** for a circular economy. These R-words are actually each strategies in their own right. From highest priority in the circular economy to lowest, the 10Rs are as follows:

- Refuse
- Rethink
- Reduce
- Reuse
- Repair
- Refurbish
- Remanufacture
- Repurpose
- Recycle
- Recover

These strategies can be grouped into three categories based on when they are implemented: during the product design, consumption, or return phase, as illustrated below. That said, as you review each of them, please keep in mind that despite the hierarchy presented, optimization is case-specific and you must consider an entire, system-wide process to make the most effective decisions. Many of the most impactful efforts use multiple R strategies in conjunction with one another, compounding their power.

10 R's
STRATAGIES FOR CIRCULARITY

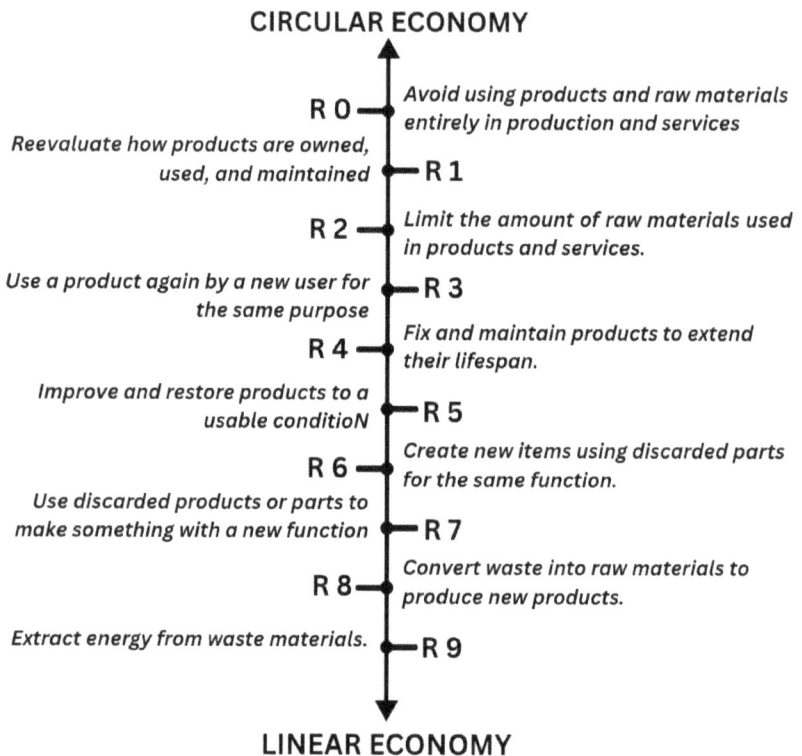

CIRCULAR ECONOMY

R 0 — *Avoid using products and raw materials entirely in production and services*

Reevaluate how products are owned, used, and maintained — R 1

R 2 — *Limit the amount of raw materials used in products and services.*

Use a product again by a new user for the same purpose — R 3

R 4 — *Fix and maintain products to extend their lifespan.*

Improve and restore products to a usable conditioN — R 5

R 6 — *Create new items using discarded parts for the same function.*

Use discarded products or parts to make something with a new function — R 7

R 8 — *Convert waste into raw materials to produce new products.*

Extract energy from waste materials. — R 9

LINEAR ECONOMY

Figure 2.1

Design-phase R strategies

Introducing circular strategies is most sustainable as part of the design phase, because it is the phase in which feedback loops are the shortest. In fact, if applied extensively, they would completely eliminate waste at the beginning of the development cycle. Furthermore, such efforts usually add value by focusing on smarter, more responsible product use and manufacturing processes. The three Rs that fall under the design phase umbrella are **refuse**, **rethink**, and **reduce**.

For businesses, to **refuse** is to ban the use of harmful materials, prevent manufacturers from using certain production processes, or to use exclusively alternative products or materials in place of those that carry a strong environmental load. An example of the latter comes from Finnish battery technology company Broadbit. They've created a sodium-based battery that lasts longer and is less expensive than the batteries that are commonly used today, while also using an abundant, earth-friendly material—as opposed to requiring mining for scarce metals like lithium, cobalt, nickel, and copper.

In certain instances of refusal, an entire product may be phased out. Policy can help lead this charge. See the European Union's ban of the 10 most common single-use plastics[40], or the United States' ban[41] of the manufacturing and sale of common incandescent lightbulbs starting on August 1, 2023.[42]

On the consumer end, this strategy can be thought of as simply **refusing** to buy products that one doesn't need. Slower fashion—buying clothes with an eye toward longevity, as opposed to changing out one's wardrobe with each passing trend—comes to mind.

Rethinking involves changing your mindset about product usefulness and ownership. The sharing economy has become a hugely popular way of rethinking how a product functions in the value chain. Successful businesses have been built around renting everything from home furniture to tech equipment, evening wear for formal events to e-bikes and e-scooters.

To **reduce** is, understandably, to decrease the use of raw materials. This can mean increasing efficiencies and decreasing waste in manufacturing. Designing for longevity also has the ultimate downstream goal of reduction, because keeping a product in use longer means putting off or even eliminating the need for a replacement, which would require the use of raw materials. A made-to-order business model is one way to reduce energy use, by eliminating the need to hold inventory as well as ensuring there's no excess product to unload.

Consumption-phase R strategies

R strategies in the consumption phase are focused on optimal use—preserving and extending the lifespan of a product and its component parts. These types of strategies are considered "medium-loop," the next best solutions in cases when short-loop, design-phase solutions simply cannot be applied. There are 5 Rs that make up the consumption phase: **reuse**, **repair**, **refurbish**, **remanufacture**, and **repurpose**.

A product is eligible for **reuse** when it is discarded by its first owner but is still in good enough condition that secondary use by another owner is possible. Vintage shops, secondhand stores, and antique markets traditionally capitalize on reuse. Patagonia's "Worn Wear" and lululemon's "Like New" programs demonstrate that major retailers are starting to play in this space. In the informal economy, garage and estate sales and clothing swaps are excellent examples of personal re-use. As the saying goes, one person's trash is another person's treasure. Another way to engage in reuse is to choose reusable products in place of single-use options: reusable coffee mugs or thermoses in the place of single-use cups, metal straws instead of plastic, and so on. Soda drinkers may be familiar with SodaStream, device technology which cuts down on plastic bottles by allowing fans of sparkling drinks to make their own bubbly beverages at home.

As is likely clear by now—and as the name makes evident—our entire business model at REUZEit is a reuse model. Our track record of putting items back into service equates to millions of dollars saved or generated, depending how you look at it. Since we gained the ability to start tracking our impact in 2014, we've enabled our clients to either reuse themselves, or to sell to another company for reuse, 1,363.62 metric tons of equipment. This amounts to about $7.6 million in total cost avoidance.

To **repair** is to maintain and fix up products for their originally intended use(s). Easier said than done, in some cases. And sometimes, organizations feel compelled to impose harsh guidelines around the practice. Before they reached an agreement with the American Farm Bureau Federation (AFBF) in early 2023, John Deere required that farmers use only authorized parts and service facilities when repairing their tractors and other equipment; this meant that attempts at self-repair, or working with cheaper, independent repair providers, would render their warrantees useless. And with Deere's integration of increasingly complex, high-tech computer systems in their equipment, fixing a tractor had become an expensive endeavor. More and more often, policy is stepping in to help promote quality repair in such industry cases. Many consumer groups are pushing for Right to Repair laws in the United States, giving consumers the power to repair their own products. Four states passed right to repair legislation in 2023[43]: Colorado, California, New York, and Minnesota.

Refurbishing refers to restoring an old product such that you improve it, or bring it up-to-date for current use conditions. This can include updating parts, or exterior, aesthetic updates (i.e., new paint). Patagonia's Worn Wear was mentioned above as an online secondhand market; the company also offers product repair services on most items for the cost of round-trip shipping.[44] Apple, Dell, HP, Samsung, and more offer refurbished product programs for their electronics, and the Amazon Renewed program offers a platform for customers to shop pre-owned, refurbished products "backed by the Amazon Renewed Guarantee."[45]

Remanufacturing involves making new products from discarded components, by integrating multiple parts from separate sources into a product that uses those parts with the same function for which they'd initially been designed and used. International automotive supplier ZF earned Cradle to Cradle certification for their remanufacturing concept: at just one of their 162 worldwide locations across 31 countries, ZF receives about 40 to 50 tons of parts daily, and they're able to remanufacture 80 to 95% of that material, putting it back into the market. They mostly supply products and systems for passenger cars but also serve the industries of wind power, construction, agricultural machinery, marine propulsion, and more. As ZF's €46.6 billion (US$51 billion) in sales in fiscal year 2023 clearly demonstrates, remanufacturing is not just a strategy for small businesses.

The last consumption-level R is **repurpose.** Repurposing is similar to remanufacturing, with the big delineation being that with repurposing, discarded products or parts are used to create a *new* product with a *different* purpose than the original design. So much innovation exists within the realm of repurposing, resulting in unique products. While some of this work exists formally, much of it lives in the informal economy. Think of broken musical instruments that have been turned into lamps or side tables at a craft fair, or a neighbor's beloved tire swing. One of the times we were working to clear out a facility for a client, an interior decorator found out what we were doing and took a couple of large, unused refrigerator cases with her. She then used them as a coat closet in a tool barn that she was converting into an industrial-themed Airbnb guest house. A smart, creative solution that cost her nothing more than the cost of gas to pick it up, and it had a great story baked in.

Return-phase R strategies

Last—and, yes, least—are the R strategies that fall within the return, or end-of-life, phase of a product's journey. These are the longest loops within the circular model, because in dealing with creative material application, they require resource-intensive processes to facilitate. The goals of return-phase strategies are to capture and retain value and use waste as a resource. The final two R strategies, therefore, are **recycle** and **recover**.

Recycling involves processing waste into either new products or new materials that can be used in the creation of new products. Recycling is of course important and necessary. But for how much emphasis society—and governments—place on it, it's key to note that it's the penultimate priority in this list of 10Rs. Some leading thinkers in the field of CE have even referred to the recycling economy as a stopgap between the linear economy and the circular economy. See the below graphic inspired by Claire Potter's 2021 book *Welcome to the Circular Economy: The Next Step in Sustainable Living*. Potter suggests that today, we're more-or-less living in the recycling economy, which is an improvement upon the linear economy, to be sure, but still results in waste—and, as we'd argue, missed opportunities for capturing value in the form of revenue.

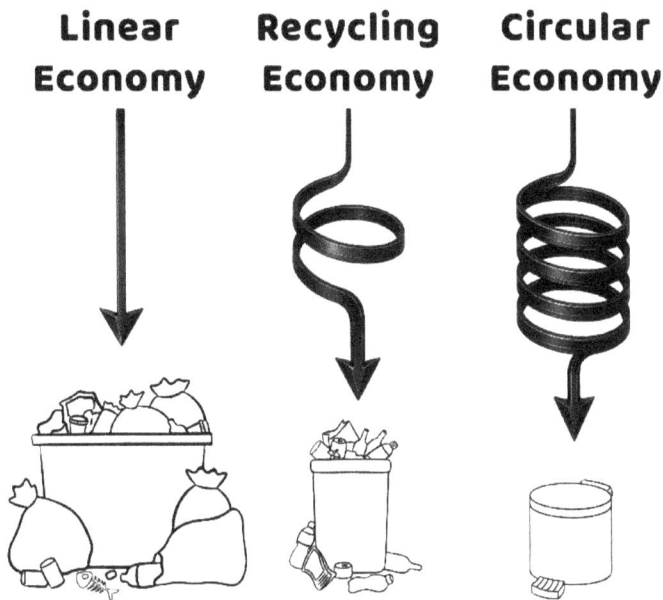

Figure 2.2

Within the umbrella of recycling, some refer to the theoretical process of *upcycling*—making relatively higher-quality materials from waste—as opposed to *downcycling*, which results in lower-value end-products. But part of why *all* recycling results in lower-value products compared to their original is that, unfortunately, in addition to drawing on resources and energy to process materials, *most recycling is essentially downcycling*. Many materials lose their durability during the recycling process, resulting in a cheaper, weaker product. Plastic bottles may be turned into carpet fibers, glass jars into fiberglass insulation, writing paper into paper towel or egg cartons, but in each case, recycled product B is weaker than original product A.

Still, two creative companies in Japan are focused on upcycling. One is Kitafuku[46], a startup that partners with craft brewers to collect their discarded malt and turn them into a sturdy, brown-tinged paper, perfect for coasters and notecards. The other is Cacao Husk Incense, which, as the name suggests, upcycles cacao husks (waste that's a byproduct of chocolate production) into incense. (The parent company, Chocolate Design Company, also uses these husks to create tumblers, as well as establishes partnerships with breweries that use the husks to create other craft beers.)[47]

On a larger scale, and considering the concept of recycling as a whole, the private, US-based recycling business TerraCycle is a fantastic case study of an organization enabling the recycling loop. TerraCycle works in 21 countries worldwide to recycle waste that local facilities don't accept, because it's not profitable to do so, though much of it can still be diverted from landfills.

If nothing else can be done with a material, the last resort is to process it to **recover** energy. (This is also known as *waste-to-energy conversion*.) As a common example, composting recovers energy through anaerobic digestion, producing biogas that can then generate heat or electricity. Biogas companies across the world use manure, food waste, wastewater, byproducts from alcoholic beverage production, municipal solid waste, and more to create energy that can be used to power residential and commercial spaces.

Framework #2: The Circular Inputs Strategy

The **circular inputs strategy** refers to replacing linear materials with those that have been or can be recycled. Think of this as, essentially, slotting circular materials into the linear model, with the big benefit being a reduction in the need for virgin raw materials. Most likely, you can recall seeing a product that's labeled as being made with X% recycled material. Tons of international companies are now taking steps toward using more recycled material in their packaging, including Unilever, Nestlé, L'Oréal, and Johnson & Johnson.[48] Or perhaps you recognize the concept of a core unit from having purchased a refurbished alternator for your car? Other circular inputs include renewable power sources or compostable products.

As viewed from a business asset perspective, when it comes to industry, the most frequently applied use of the circular inputs strategy is when manufacturers re-use material or products in a new model, or for service support. This material could be from excess or obsolete stock or could be pulled from the supply for a discontinued model.

At REUZEit, our excess & obsolete (E&O) clear-out work supports the circular inputs strategies of other companies. Many companies earmark their E&O materials for the landfill, or, at best, mixed recycling. In one instance, our client gave us a disorganized, vague list of all their E&O materials to deal with. There were truckloads of stuff. But when we began to process all the valuable things, it turned into an impressively robust list: motors, pumps, long coils of copper, tools, power supplies, heat exchangers, and more. We listed each worthwhile item on our eBay store at the time, "Lab Treasures," where they were snatched up by service companies, innovators, and inventors who wanted to use these parts for new things as inputs in their own creation processes. For some items, we also sought out contractors we knew would benefit from materials we had on hand—a method we frequently employ. In the above illustration, we knew that refrigeration contractors, who service components, would find value in the pumps, among other items. They took us up on the offer, purchasing things they needed at a fraction of the price of buying new, both increasing their profitability and limiting their need for raw material use.

We also often try to reuse materials in our own operations. Most of our furniture, artwork, and even plants in our REUZEit offices are from site shutdowns. One time, we were closing a facility, and no one wanted about 200 indoor potted plants and trees. We did our best to sell some, but we had to give away most of them. Some employees and businesses in our area got a small jungle's worth of houseplants for free.

In a circular input context, another lens through which to consider the impact of reusing is that of carbon equivalents (CO2e). With REUZEit's reuse model, the only carbon "cost" is the energy needed to move an item from one place to the next—which is next-to-nothing when considering all the carbon needed to power the production, packaging, and distribution of a new item. In fact, we found that reusing is 99.7% more effective at reducing carbon as compared to buying new. The carbon equivalents saved by our efforts equate to:

- Enough electricity to power 58,455 households for one year—more than all the housing units in the city of Lansing, Michigan.[49]
- The removal of 69,987 passenger cars from roads for one year—more than all cars in the city of Ithaca, New York.
- 394,273 acres of US forest growth in one year—only 10,000 acres shy of the size of California's Sequoia National Park.[50]
- 5,321,180 tree seedlings grown over the course of ten years. We aren't sure how to help you imagine over 5 million trees, but let's all agree—it's a lot of trees.

REUZEit is growing, but we're still just one company. Part of what drives us, and why we're trying to share this message about the circular economy, is how incredibly these numbers will scale even further, when we get even more people onboard. Today, pretty much everyone agrees that recycling is easy and that it makes sense—this wasn't the case in the '90s, but society made a shift. So, if we could move that mindset up the value chain and help people understand that re-using is just as easy as recycling—organizations like ours make it so!—far more lucrative, *and* better for the environment? Just imagine how much carbon consumption could be reduced. We have the chance to set the future up for success by establishing the reuse economy now. Existing circular input strategies at existing companies are just one avenue for growth.

Framework #3: The Sharing Economy Strategy

We talked about the **sharing economy** in framework #1, as part of the R strategy of rethinking; however, it merits another overview here. The sharing economy is the practice of borrowing or renting products, frequently organized and systemized using online platforms. It's also known as *collaborative consumption*. The term *sharing economy* is often used broadly, as an umbrella that can encompass a wide range of products and services.

Critically, the sharing economy is nothing new. Concepts like carpooling and community gardens (or allotments, in England) have been around for ages. As has the idea of trading skills, services, or goods—like helping a friend with a leaky sink in exchange for some choice clippings from their garden. Or pooling skills, as with the tradition of barn raising in Amish and Mennonite communities. Nevertheless, the circular economy is emerging as a targeted extension of this societally timeworn practice, and entrepreneurs are codifying it— that is, profiting from the platforms and apps they design and the networks they're able to grow.

We were on a tight budget when we first started REU-ZEit. We found this new platform called RelayRides in San Francisco that allowed us to rent out our vans or pickup trucks when we flew to the Bay Area to work on projects, and we met a lot of cool, interesting, like-minded people who were also making money by renting out their cars on the platform. (Many of them left their cars at the airport to be rented out while they were traveling, so they got free airport parking in addition to making income from a car that would otherwise be sitting in a driveway or parking garage.) RelayRides eventually became Turo, the world's largest car sharing marketplace with over 370,000 active listings as of summer 2024. They raised a $250 million Series E in 2019, then giving them a $1.25 billion valuation, with revenues of nearly $900 million in 2023[51] and a current valuation estimated around $2.7 billion.[52] There's obviously some solid money to be made in this sharing arena, and innovation opportunities are nearly limitless.

Framework #4: The Product-as-a-Service Strategy

Product-as-a-service (PaaS) refers to company owners maintaining ownership of a product even after it's "sold," making them responsible for its function and maintenance—and, ultimately, its disposal. (We've mentioned this concept before, too, when we discussed the CE benefit of increased demand for services in chapter 1.) This shift, from selling physical products to selling functions and benefits, puts the onus on and motivates product designers, perhaps even more than the consumer or user, to optimize durability and longevity from the get-go. In PaaS scenarios, product disposal is not only a cost for the company to bear but also means that said product will no longer be bringing in revenue—so it hits both sides of the balance sheet. In other words, the business incentive is not just on the front end of the value chain, as in the linear model, when companies work to maximize units sold to maximize revenue. You might be thinking that this makes sense in theory, but how do companies make money if they're not charging for the products themselves? The most frequently used PaaS revenue models are leasing or pay-per-month—i.e., subscription, pay-per-use, and/or pay-for-performance.

An example of PaaS with which you are likely intimately familiar, but that you might not have considered to be circular, is your internet router. It's probably from your internet service provider. Many of us rent our router, paying for its use as part of our monthly bill. Wondering what happens with your router when you close your account? Drake & Farrell in Blieswijk, Netherlands, one of our neighbors in Europe, has built a business around a circular supply chain for e-waste like this. They collect routers from people who close accounts (reverse logistics), then perform inspections and refurbishments, then send the internet routers back out to new clients (forward logistics).

[A quick note on terminology here: In the academic world, product-as-a-service is also known as Product-Service Systems (PSS). However, in the broader business world, the acronym PaaS may also be used to

refer to *platform*-as-a-service; that's the cloud computing service model that you often see in the tech world,
where users access software or applications, paying
for packages and/or subscriptions. Think Salesforce,
Google App Engine, or Microsoft Azure. For the purposes of the circular economy, PaaS refers to product-
as-a-service.]

Framework #5: The Product Life Extension Strat-
egy

This strategy is well-named. It involves prolonging the
lifespan, utilization, and ultimate value of products
through repair, remanufacture, and resale.

The big benefits of product life extension are threefold:
1) reducing the demand for new production, and therefore the need for raw material extraction, 2) reducing
energy consumption and carbon emissions involved in
the process of extracting raw materials, and 3) waste
reduction (or less stuff in landfills). More broadly, product life extension also cuts down on embodied emissions, or the carbon gas emissions created during a
product's lifecycle, from extraction of resources, production, use, and disposal.

One important caveat: always think holistically. The goal with the product life extension strategy is to *optimize* a product's lifespan, not necessarily to *maximize* it. Using more resources to create a product that will outlive its use is more wasteful in the long run. The best-case scenario is for a product to last as long as a consumer will use it. Careful planning and concerted efforts toward understanding customer use are key. As an illustration, designing a power drill for hours upon hours of continuous use simply doesn't make sense. Even the busiest contractor will only use a power drill for so long before they're done for the day, and often with many stops and starts throughout their project work.

Canadian office furniture company Envirotech uses a product life extension strategy to improve sustainability outcomes for a wide range of customers. Their office furniture remanufacturing program extends the life of resources like wood, steel, aluminum, plastic, and fiber. Notably, the Envirotech team found that office lease durations were shrinking[53] and, with less than 10% of office furniture being repurposed beyond its initial location, much of it was ending up in the landfill. They remanufactured over 40,000 workstations/desks, with limited lifetime warranties, and 100,000 chairs, with 10-year warranties, diverting an estimated 11 million kilograms (over 24 million pounds) of furniture from landfills.

Here, we'd like to acknowledge that while these concepts might be nothing new to some at this point, it can be helpful to have different ways of approaching CE. If transitioning to the circular economy feels overwhelming, focusing your efforts on product life extension, for example, can help you think about CE in a way that feels manageable, and doable. If you remember nothing else from this book beyond thinking about ways you can extend a product's lifecycle, we'd still consider that a success.

Framework #6: The Design-for-Recycling Strategy

The concept of **designing-for-recycling** is to go beyond designing with recycled materials in mind to selecting materials *based on* their recycling rate. This means using fewer dissimilar materials in products, and/or allowing for easy disassembly of components made from different materials, to improve the potential for material life extension. German association Verein

Deutscher Ingenieure (VDI) recommends thinking about recycling during three stages: production, use, and after use.

See the table below for specific example design-for-recycling product guidelines[54]:

Area	Guidelines
Materials	i. Minimize the number of different types of materials ii. Make subassemblies and inseparably connected parts from the same or a compatible material iii. Avoid the mixing of materials in assemblies iv. Mark all plastic and similar parts for ease of identification v. Use materials which can be recycled vi. Ensure compatibility of ink where printing is required on plastic parts vii. Avoid composite materials viii. Eliminate incompatible labels on plastic parts

	ix. Hazardous parts should be clearly marked and easily removed
Fasteners and connection	i. Minimize the number of fasteners
	ii. Minimize the number of fastener removal tools needed
	iii. Fasteners should be easy to remove
	iv. Fastening points should be easy to access
	v. Snap-fits should be obviously located and able to be disassembled using standard tools
	vi. Try to use fasteners of material compatible with the parts connected
	vii. If two parts cannot be compatible, make them easy to separate
	viii. Eliminate adhesives unless compatible with both joined parts
	ix. Minimize the number and length of interconnecting wires or cables used
	x. Connections can be designed to break as an alternative to removing fasteners
Product structure	i. Minimize the number of parts
	ii. Make designs as modular as possible, with separation of functions

	iii. Locate unrecyclable parts in one area which can be quickly removed and discarded
	iv. Locate parts with the highest value in easily accessible places
	v. Design parts for stability during disassembly
	vi. Avoid molded-in metal inserts or reinforcements in plastic parts
	vii. Access and break points should be made obvious

Or, more broadly, the government of Catalonia's Catalan Centre for Recycling[55] suggests a company follow these six steps, which we've rephrased a bit, to incorporate design-for-recycling into its products:

1. Commit to change, organization-wide
2. Create a team and develop a schedule
3. Analyze the product(s)
4. Assess the environmental impact of the product and propose solutions
5. Prioritize improvements based on impact, create a plan for their introduction, and introduce them

6. Assess the effectiveness of the improvements and set up follow-ups focused on continuous improvement

One organization enabling design-for-recycling efforts is the French biochemistry organization Carbios. They've developed enzymes that can break down certain polyesters, in particular PET (polyethylene terephthalate) and PLA (polylactic acid), a biobased polymer. Therefore, if manufacturers can use these materials in a way that they can be easily processed on their own at a product's end-of-life, they go through an enzymatic recycling process that results in the production of biodegradable plastics—which get fed straight back into the loop.

Framework #7: The Zero-Waste Manufacturing Strategy

REUZEit helps our clients achieve zero waste. That is, we will do everything we can to find a new home for equipment. (Drawing back on framework #1, a robust circular asset management program can ensure all equipment follows the 10Rs so that every effort is made to reduce waste.) We've donated equipment to afterschool robotic clubs and artists. We've sold materials to decorators who want to save money but also want to show their clients the value in reuse.

One fun experience we had was supplying the lab equipment for the movie *Jurassic World*. The producers approached our client, Thermofisher, to pay them for use of their equipment as props in the movie's lab set. And once they heard about our involvement, they also asked if they could have some of our old, used equipment so that it could be ransacked by CGI dinosaurs. After the crew walked through our warehouse, we labeled everything they were interested in and sent out two trucks full of lab equipment. The coolest part? We were invited to be on set so that we could set up all the equipment in the secret lab. We even got to help make the robots move! It was super fun. And the producers were great about handling the equipment, so most of it ended up coming back to us to be resold, too.

Additional Targeted Strategies

There are a few more strategies we'd like to cover here that are less widely applicable but nevertheless valuable to consider: extended producer responsibility

(EPR), regenerative agriculture, and biomimicry.

Extended producer responsibility (EPR) is largely driven by legislation rather than by corporate policies or initiatives. The general idea is to add all the estimated environmental costs associated with a product throughout its lifecycle to the market price of that product, usually applied at a product's end-of-life. Financial incentives encourage manufacturers to design environmentally friendly products as well as to offer take-back, recycling, and disposal programs.

In many countries, tires are subject to EPR. (Tire EPR has been in place in Europe since 1995 and in Canada since 2007[56].) According to the Product Stewardship Institute[57], some nations are able to reuse, retread, and recycle 100% of their tires. This is massively impactful, as retreading one tire saves 40 pounds of materials. In the US, tire EPR bills have been introduced in both Connecticut and Vermont, the former of which passed a first-in-the-nation tire EPR bill in 2023.

We'll come back to a broader discussion of worldwide standards and regulations in chapter 7.

Next, **regenerative agriculture** is a farming method that focuses on restoring soil and ecosystem health. The six core principles of regenerative agriculture[58] are as follows:

1. Understand context
2. Minimize soil disturbance
3. Diversify crops
4. Protect soil surface
5. Maintain living roots
6. Integrate livestock

(Understanding the context of your operation is the most recent addition to this list and speaks to both the comprehensive nature of this strategy as well as the importance of adapting it for the needs of individual sites.)

While these tenets are specific to agricultural products, they're also deeply connected to the circular economy because of their promotion of the efficient use of natural resources. The Natural Resources Defense Council (NRDC)[59] calls it a dynamic system and philosophy built on holistic principles that ask us "to think about *how all aspects of agriculture are connected through a web*—a network of entities who grow, enhance,

exchange, distribute, and consume goods and services—instead of a linear supply chain." In addition to
restoration, regenerative agriculture aims to "address
inequity, and leave our land, waters, and environment
in better shape for future generations." Specific practices vary among growers and regions. Many Indige-
nous communities have long followed these
philosophies, though likely without use of the "regenerative agriculture" label, for generations upon generations.

6 Core Principles of
REGENERATIVE AGRICULTURE

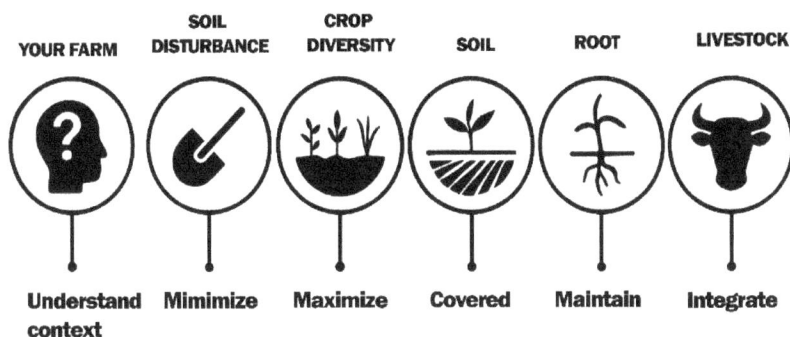

| YOUR FARM | SOIL DISTURBANCE | CROP DIVERSITY | SOIL | ROOT | LIVESTOCK |

| Understand context | Mimimize | Maximize | Covered | Maintain | Integrate |

Figure 2.3

Examples of worldwide regenerative agriculture include the following: grassland management programs
that engage with local pastoralists in Northern

107

Tanzania;[60] sheep grazing amongst the grape vines at vineyards like Tablas Creek in Paso Robles, California[61]; farms in Iowa using the prairie strip planning method to protect soil and water, while providing a habitat for wildlife[62]; and tea farms in Sri Lanka that committed to herbicide-free weed management to protect their soil health and saw a crop yield increase of up to 20% by the third year.[63]

Finally, **biomimicry** is the practice of learning from and mimicking nature to solve human problems. The idea is that if birds can fly without fossil fuels, plants can use photosynthesis to get all the energy they need from resources that exist in nature, and ecosystems can sustain themselves—basically, life has figured it out—so we can, and should, learn from it. Biomimicry came to the forefront with Janine M. Benyus's book by the same name, published in 1997.[64] She founded the Biomimicry Institute in 2005. The idea is credited for being one of the major original inspirations for circular economy thinking.

One of the most famous examples of biomimicry is Velcro. Swiss engineer George de Mestral[65] came up with the idea for it when he got back from a walk with his dog, both of them covered in burrs. He looked at the burrs under a microscope and noted the tiny hooks at the end of their spines, which allowed them to cling easily to anything with a loop, like his dog's hair and the fabric of his pants. De Mestral then replicated this simple design synthetically, with a two-part system, one a strip of tiny hooks and the other a strip of tiny loops.

It took de Mestral a decade to go from his idea for Velcro to creating it, and another four years to get it patented. The road to circularity isn't always quick and easy—it can be a challenge. It requires you to take a big step back and rethink both what is and what can be. The good news? Awareness of all these strategies and frameworks of thought is the first step in helping you make the leap. We hope you found a strategy or two that resonated for you. Perhaps you've even been inspired to begin building a CE roadmap for your organization. Let us help you with that...

Get Started: Exploring Circular Economy Business Models

We love a guide from Twice Commerce[66] that offers an 8-step process for creating a circular economy strategy for consumer businesses. It allows you to design a model that suits your individual and/or organizational needs, ideal for business owners and entrepreneurs. We encourage you to interact with the guide and explore the models within that might work for you and your company.

Next, a useful Stanford Social Innovation Review article called "Designing Your Circular Business Model"[67] breaks things down in terms of four loops:

- **Narrowing resource loops:** Using fewer resources, to make the production process more efficient.
- **Slowing resource loops:** Extending a product's life to reduce consumption.
- **Closing resource loops:** Reusing material after it's been in the hands of the consumer (i.e., post-consumer recycling).
- **Regenerating resource loops:** Improving the environment that you exploit for operational and commercial use.

Review the abovementioned article for more information on the desirability, feasibility, and viability of each of the above loops, to help you determine which stages of a product can best be improved within your current system.

At this point, we'd like to admit that we kind of lied to you earlier. There is one more CE strategy we want to explore—a big one, that is close to our hearts and deserves particular attention. After all, we've dedicated many years of our lives to it. Join us in the next chapter as we delve into the world of equipment recovery.

Chapter 3

Equipment Recovery: Reaping the Benefits

Once you've employed as many CE strategies as feasible within your organization's wheelhouse, it's likely that you'll still have excess inventory and/or assets that you no longer have use for. As tempting as it may be to jump ship—or try to get out of the loop, as it were—this is actually where things can get exciting. We know, we used the word "exciting" in a chapter about equipment recovery. But, by the end of this chapter, we think you'll be excited about the potential benefits for your business, too.

What Is Resource Recovery?

To start, you'll need to more deeply understand the broader process of resource recovery. In an organizational context, the concept of resource recovery acknowledges that, despite efforts to implement change in the design phase with the three most valuable of the 10Rs (refuse, rethink, and reduce), companies still often find themselves with excess resources in the consumption or end-of-life phases. This excess can be in the form of unused raw materials, waste resulting from the production process, outdated equipment, or other goods that are, simply, at the end of their first lifecycle. When further *reduction* of waste isn't possible, the *recovery* process extracts all recoverable materials to reduce the amount of waste going to the landfill.

A resource recovery model

The goal with resource recovery is to reuse waste by converting it into a resource. To implement this model, organizations must first conduct an in-depth analysis of how to recover all waste and byproducts that result from each step of their production process.

Strictly from a bottom-line perspective, the goal is to reduce or eliminate the costs associated with both disposing of old resources and obtaining new raw materials to replace them. To that end, the three guideposts that lead the way include the following:

1. Extract more value from existing resources.
2. Create less dependence on raw materials.

3. Identify revenue opportunities from waste or by-products.

Importantly, items 1 and 3 both offer the chance to develop new revenue streams within existing production processes, while item 2 cuts down on operational costs.

Applications in the real world

International engine and equipment manufacturing corporation Caterpillar has developed a system to extract more value from a critical resource used in their production processes: iron. By charging customers a deposit that is refunded when they return their used components, Caterpillar increases their recovery rate of end-of-life equipment. They also have a system to manage product returns from dealers and inspection facilities. Because they've built these systems into their product lifecycle, they're able to salvage 70% of their newly designed machinery and recycle more than 150 million pounds of end-of-use iron each year.[68]

In a general sense, large industrial businesses can create less dependence on raw materials by reducing their operational use of fossil fuels. This can be achieved by recovering heat generated from product destruction and combustion and using that energy to fuel the creation of *new* materials. As another industry example, food and agricultural waste can be converted into bio-methane gas, which can then be used to create electricity or hydrogen.

At a grassroots level, in *The Circular Economy in the Global South,* chapter author Priliantina Bebasari looked at women's empowerment groups in the greater Jakarta area that had identified revenue opportunities for low-income and disadvantaged women using waste or byproducts in upcycling enterprises. One such group, PEKKA (roughly translating to Women-Headed Family Empowerment), organizes over 2,500 women's groups with members who have become heads of their household. Local governments led trainings for these groups to learn strategies for proper waste management and how to create valuable products out of waste. The women upcycled plastic waste into bags and tablecloths and textile waste into doormats and blankets, which they then sold as part of a group enterprise.[69]

Benefits of resource recovery

There are potentially impactful financial advantages to the resource recovery process. It's no surprise that reducing raw material use cuts down on costs, increasing profit margins. But moreover, efficient and cost-effective methods of reusing and recycling waste can allow organizations to recover resources at values equivalent to or even above initial investments, either cancelling out a line item from the operational costs ledger or adding one to the list of revenue streams.

Of course, the other significant benefit of resource recovery is environmental. Whether your company's goal is to meet internal or external sustainability targets, or share your story with environmentally conscious consumers, or even to use such initiatives to attract top talent, your efforts will reward you with powerful examples of environmental impact.

How About Circular Asset Management? The REUZEit Approach

We started REUZEit over 15 years ago using the concept of circular asset management to buy and sell used lab equipment. We've built and scaled the company around rethinking how corporations and laboratories approach and manage the tools that make their work possible. In these last two-plus decades, we've helped our clients, mostly within the biotech and pharmaceutical industry, recapture millions of dollars from their surplus assets while simultaneously keeping literal tons of used equipment out of landfills. Our team specializes in managing, refurbishing, and reselling surplus lab equipment for some of the world's largest scientific producers, making us uniquely qualified to work with high-quality equipment.

Within the circular economy, Surplus Asset Management (SAM) provides a solution for optimizing the use of resources, assets, and equipment throughout their life cycle, thus enabling new business models, increasing profitability, and achieving sustainability goals—all while reducing obsolescence and waste. In a nutshell, Circular Economic Surplus Asset Management (CE–SAM), or, as we call it, Circular Asset Management (CAM), is a straightforward, viable business approach that creates value by extending the lifecycles of would-be discarded but-still-useful equipment.

The troubling reality of equipment waste should be of significant concern for all organizations, not only for its

detrimental environmental impact but also for its negative financial repercussions. While the significant financial liability that excess assets carry for biotech,
pharma, and life sciences labs shouldn't be over-
looked, we often find these firms are leaving hundreds
of thousands, if not millions, of dollars on the table. As
we say to our customers all the time, excess lab assets
are more costly than you might think, even when
they're just sitting there. Such costs are associated
with the following business liabilities:

- Redundant equipment procurement
- Unnecessary equipment storage
- Unused equipment
- Below-market equipment resale

In the face of such perceived obstacles, so many organizations choose to cut their losses—i.e., throw
equipment away. Every month, in every country, literal
tons of unused equipment are sent to landfills worldwide. The scale of this waste is massive. While, in
many instances, the lifespan of this equipment could
have been extended several times over, it ends up de-
stroyed—a loss in and of itself. And then it causes toxic
materials to pollute the environment, as most of this

equipment will either not decompose for generations or be incinerated, resulting in toxins being released into the atmosphere. This, in turn, causes the following issues:

- Toxic metal contamination
- Chemical pollution
- Imperishable material waste in landfills
- Groundwater contamination

As we mentioned in the last chapter, our research[70] shows that keeping a piece of equipment in use, either by redeploying it within the company or selling it on the used market, reduces that piece of equipment's carbon legacy by 99.7%.

At REUZEit, our CAM efforts result in unheard of financial and environmental successes for our customers. We offer capital investment recovery at a net zero cost. These results are made possible through our digital dashboard, which is the world's most advanced circular economic platform within our industries of specialization. (Aside from providing transparency to decisionmakers, our dashboard connects users to our comprehensive suite of onsite services that you can't find anywhere else.) Our worldwide network then enables us to pick up, ship, store, and refurbish lab equipment anywhere, anytime.

Let's make this tangible for you with another case study. Over the course of five years, we worked with a leading life sciences company with more than 1,000 sites and 75,000 employees across the world. In this time, we developed an international equipment management system and streamlined process, with real-time visibility and reporting, where before there had been nothing. Our end-to-end equipment specialists were on hand for this client, reducing the internal resources they needed to commit to the process. That all sounds nice enough, but the real proof of success is in the numbers:

- 502% ROI for the program
- $15,000 average cost avoidance per redeployed asset
- 40% savings on services rendered
- 4.6 million pounds of equipment diverted from landfills
- 7-figure annual donation initiative

(While these highlights represent success at a large company, the first three bullet points could be hugely impactful no matter the size of your organization. And for those large clients, who can't usually find turnkey, bolt-on solutions like ours—especially not those that so easily pay for themselves—we've been in meetings with C-suite executives who've called our CAM platform a "no-brainer.")

Beyond these bottom-line impacts are tangible environmental wins—benefits that often go hand in hand, as across industries, environmental mindfulness has come full circle. In the early days, it took the form of activism—the people demanding that big industry do something to stop waste and pollution. Today, there are thousands of major companies across the world that have made the paradigm shift in perspective on sustainability, understanding that we are all stewards of our environment. Businesses are making strides not only because of external demands, but because the people who work there are creating committees and starting internal programs to make their organizations more sustainable.

Most of the employees of clients that we meet tell us about the mountains of valuable equipment they had to throw away or recycle the last time they cleared out their storage closets. After they find out that their organization has partnered with us, granting access to the REUZEit CAM platform, they are thrilled. No more feelings of helplessness, no more guilty consciences. At all levels of the organization, employees can just click a few buttons to request an equipment pickup. The equipment owner can designate if they want REUZEit to hold the piece of equipment, sell it, or make it available for anyone in their company to use. And for owners who want their equipment donated, we host a private donation portal where nonprofits can access top-quality equipment for free.

Long before REUZEit had our cutting-edge dashboard and donation portal set up, we were facilitating equipment donations. At one point, we received two truckloads of unused microscopes to sort through and attempt to redeploy. We managed to sell most of the high-end microscopes to universities and researchers over the course of a year, recovering more than $250,000, but there were still thousands of basic microscopes for which we had to find a market. These were basic and unbranded, in unmarked packaging—and they'd never been used. A few microbreweries and wineries bought them for basic yeast and bacteria analysis. It hardly made a dent in our inventory. Then, we had a eureka moment. The basic microscopes were perfect for primary and secondary education. On behalf of our client, we facilitated a donation to the Los Angeles Unified School District valued at $750,000.

Best Practices for Equipment Lifecycle Management

We want to be clear—within the context of both the circular economy and our own business, we see redeployment as a secondary resort. Throughout our years in this field, we've developed a list of best practices that will further reduce the need for an organization to resort to asset and equipment recovery in the first place. This requires starting at the beginning of the lifecycle of each piece of equipment and includes checkpoints throughout its use.

Connect purchasing decisions to business strategies

We recommend that, before a firm even purchases a piece of equipment, they are deliberate and strategic about its potential lifecycle. When there is a robust network available for reuse—as when an organization is partnered with REUZEit—it makes sense to start by making your best effort to accurately predict the length of time for which a new piece of equipment will be needed, or in use. Another complementary tactic is to proactively identify other upcoming projects or initiatives that could make use of this new equipment. Our CAM platform can show incoming equipment to other users in the organization, including an estimated date at which the equipment will be no longer needed for its initial use and therefore ready to be used on another project or by another department. And there are built-in communications tools that make it seamless for team members to make these arrangements across departments, divisions—even other countries—or just with another lab down the hall. Beyond streamlining logistics in large companies where things may tend to get siloed, this gives employees the chance to adapt their projects and budgets from the get-go, in ways that help them take advantage of predictive reuse schedules.

Maintain a comprehensive inventory

Inventory management is straightforward when you're dealing with new items, because the condition of all the equipment is known: new, perfect, in a box, off a shelf, ready to use. When it comes to developing a successful reuse program, you also need to know the condition of equipment that's actively being used and, ideally, its history. A comprehensive inventory will include maintenance and service history records, so the technicians refurbishing the units for next use are better prepared to recondition and recertify each item. At REUZEit, often when we see a piece of equipment for the first time, we know that it's been used…and that's about it. When there isn't a record of work performed on the device, we start a service history record. Going forward, our Equipment Lifecycle Software tracks such data so that historical service and maintenance records are carried on to the next owner and the next reconditioning technician. This makes it easier for the item to be used again and again and again by cutting down on time, labor, and other resources spent on troubleshooting, diagnostics, and any potential ineffective and/or short-term fixes. (Our inventory platform also comes with an "easy button" to get any item fully reconditioned, recalibrated, and/or reinstalled, should the current owner want to maintain its use.)

Monitor for issues

Real-time performance monitoring can predict when a piece of equipment might fail. For example, in the case of a -80-degree Celsius freezer, a cheap monitor can measure and flag not only decreased temperature stability but also increased draws in current. (This matters because a gradual rise in amps used by the motor could pre-emptively signal an impending system failure.) Of course, in this instance, recovering the contents of the freezer is the first and most pressing concern. However, it gives you the chance to intervene *before* disaster strikes. And the record of readings from the monitor provides valuable diagnostic information about the freezer itself, indicating which part(s) may need to be replaced so that it can be reused.

Track day-to-day equipment use

When equipment is turned off, it's not providing value. Monitoring the daily use of equipment helps identify whether a company's assets and capital are being used efficiently. This can be useful for planning: forecasting expenses, making decisions about future equipment purchases, and more. It can also demonstrate the opportunity for potential equipment sharing—whether internally, across departments, or perhaps externally, with a partner organization or rental program—and reuse. If something isn't getting turned on enough to merit the space it's taking up or the capital it represents in potential resale value, then circular platforms make it simple for employees to add it to a pickup list or to give it to an internal CAM program for reuse or resale.

Create a procurement plan

You can inform future procurement strategies and activity by looking at your equipment usage data—things like which tools are most frequently used, which break down or fail the most, and which last the longest. That is, staff in charge of operations and purchasing can more effectively budget plan and minimize spending on reserve equipment or parts. Plus, you'll have data and leverage to use at the negotiation table for any new equipment purchases.

Create maintenance cycles

Deferred maintenance can hit your budget hard. Running equipment until it reaches the point of failure may lead to expensive repair costs, including potential fees for emergency repair services and/or overtime costs; pricey operational downtime, which can have impacts not only on your output but also on staffing and employee morale; and, often, a shorter life for the piece of equipment that fails. Automating your equipment maintenance based on usage data distributes use, wear, and tear across all your assets, remediating this potential hard hit before it happens.

Optimize equipment distribution

Monitoring usage data will also bring patterns to the fore. Noticing how and when employees use particular pieces of equipment gives you the opportunity to optimize each piece's placement on the manufacturing floor or in the lab, for efficient workflows.

Automate the procurement process

A smart, well-maintained inventory management system can make it so much easier to manage equipment upgrades. Rather than manually keeping tabs on things, integrating alerts that notify you when parts need to be reordered or capital assets are nearing their pre-determined end of life—at least with you!—automates the process. On that note, it can even help managers get a jumpstart on equipment resale. Being able to anticipate the need for equipment upgrades allows employees with budgeting responsibilities to earmark additional funds they can put toward any purchasing needs.

Apply quality controls

Quality control checklists can be built into your equipment inventory system. These lists should be specific to each asset and can be personalized based on employee roles. By keeping this QA data in the same place as your inventory, it acts as a supplemental service history and usage record.

Use the right management technology

In today's automated society, locked cabinets and spreadsheets work for an increasingly small percentage of businesses. But smart lockers and fully integrated digital platforms—like our proprietary dashboard at REUZEit—provide so much efficiency and ease that they are often more than worth the spend. Importantly, a robust management technology makes it simple to automate the best practices in this list.

Optimizing Asset Lifespan

Optimizing an asset's lifespan involves extending its useful life safely and effectively, to deliver maximum value over time. It's both an art and a science.

To estimate how long a critical asset will bring value to your organization, before it ought to be replaced, its useful life—or, how long you can reasonably expect to use it for the benefit of your company—is weighed alongside its anticipated depreciation (an accounting measure that spreads the cost of the asset over its useful life). For the sake of this chapter, we're talking about fixed assets, or long-term tangible assets (like machinery and equipment) as opposed to intangible assets (like copyright or goodwill). We haven't developed a system to reuse or refurbish goodwill quite yet—we'll leave that to the PR professionals, publicists, and perhaps spiritual gurus out there.

For the calculation described above—to determine asset life expectancy—the following criteria and data points are used:

- Manufacturer specifications
- Past experience(s) with similar depreciating assets (drawing on your data from the best practices in the previous section!)
- Standard industry practices
- Engineering estimates

[The IRS provides a capital asset useful life table in IRS 946, Appendix B[71], that you can consult. Also, asset operations management company ToolSense has an equipment management sheet template that you can download from their website[72] if you give them your email address. Both tools are great starting points for smaller and/or younger organizations at the earlier stages of building their asset lifecycle and maintenance systems.]

Now, when you can calculate a product/asset's natural lifespan, you can work to strategically extend it. Beyond the best practices mentioned in the previous section, here are a few more ways you can maximize asset longevity in your business:

1. **Train your team.** While proper training requires time and dedication of resources in the early stages, and it requires patience before putting a piece of new equipment to work, it's always worth the effort. When team members know how to operate, troubleshoot, and maintain equipment, they have both confidence and a sense of ownership and responsibility toward each tool. Note: training should be revisited upon any update in technology or change in operational procedure.

2. **Embrace technological solutions.** We previously touched on the importance of monitoring systems and inventory software. The real-time insights certain tools offer can help you coordinate a predictive maintenance schedule, avoid downtime, and reduce human error and wear and tear. While investing in the appropriate solution(s) for your organization can come with an upfront cost that may seem substantial, we've found such tools usually pay for themselves—as they ultimately result in long-term savings in the form of asset life extension and reduced maintenance costs.

3. **Invest in quality.** Similarly, investing in assets is a long-range play. While immediate budgets must always be considered, total cost of ownership is often lower with higher-quality, more expensive equipment that performs better and outlasts cheaper counterparts. Energy efficiency, maintenance costs, and extended lifespan are important factors to weigh alongside ticket price during the asset acquisition phase.

4. **Foster a repair-over-replace culture.** Get everyone on board with the "fix it first" mindset. Make sure your teams understand that repairing first not only helps the bottom line (by extending the life of equipment versus buying new) but also contributes to your company's sustainability efforts and environmental stewardship credentials. See #1 above—training initiatives are a good place to start. Taskforces, dedicated staff, competitions, and compensatory or other rewards for innovative repair solutions are ways leadership can get creative in nurturing such a culture.

5. **Foster relationships with suppliers and service providers.** We've talked a ton about

systems and tangible assets like giant freezers and heavy machinery, and we'll continue to do so throughout this book. But none of us can ever forget that, first and foremost, we're working with *people* on a day-to-day basis. Relationships in business are paramount—even more so in a circular economic model, where dependencies have an outsized influence on the system's function and success. Suppliers and service providers who know your business well can offer advice, tailored solutions, priority servicing, potential discounts, early access, and more. While the cost savings these relationships provide can't always be captured with a measure of ROI, it's often because the right people keep things from hitting the P&L in the first place.

Now, once you've done your best to manage your equipment lifecycle efficiently and optimize each asset's lifespan, it may come time to sell your assets.

Can I Sell My Own Assets?

The short answer: yes. You can sell your own surplus assets. After you've determined which assets can no longer be repaired or redeployed internally, the next step becomes figuring out what to do with the surplus. Many business leaders turn to trusted partners, like our team at REUZEit, to help them recover the maximum possible revenue from their initial investment. An experienced reverse supply chain partner has a network of marketplaces and buyers to tap into, but that doesn't mean you *have* to work with someone like us. If you were to do this yourself, there are a handful of methods we recommend for resale, as detailed below.

Unreserved online auction

An unreserved online auction functions the way a typical in-person auction does, but online. Online marketplaces list surplus assets with a countdown timer informing buyers when an auction will end. When time expires, the highest bidder wins the product, purchasing it for the amount bid. Selling via this method works best for assets that are commodities—i.e., in high demand on the secondary market. It's also quick and efficient, reducing stress and giving you, the asset owner, fast access to capital liquidity.

Reserved online auction

The big difference between reserved and unreserved online auctions is that in reserved auctions, sellers begin with a minimum acceptable price, or a starting point for all bids. Think of an auctioneer saying, "We'll start the bidding at $200, folks." This is called the *reserve price*, hence the name. If you've ever bid for an item on eBay, this is probably how it worked. Reserved online auctions are a smart choice for high-quality or valuable items with a smaller demand, to ensure a baseline return on investment.

Private sale catalog

Another method some sellers use on sites like eBay: private sale catalogs give buyers the chance to either *buy it now* or *make an offer*. Sellers set the *buy it now* price at or close to the highest end of their expected resale value, meaning if the buyer is willing to pay that price, the seller will be happy with the transaction. When no buyers go for the *buy it now* price, offers below it are presented for the seller to consider. This is a good strategy for items that you don't need to sell quickly and that may be niche: low in demand but with high value when potential buyers do come along.

Online sealed bid

Ever been to a fundraising event where you placed a private bid on something? Perhaps you put it in an envelope near the item you were bidding on? This is a traditional sealed bid auction, and the online format is modelled after it. All offers must be submitted before a predetermined deadline, and nobody knows what other offers are like as they submit their own. This sense of mystery and competition can drive the value of bids up. Once the deadline passes, the bids are presented to the seller for approval. This type of transaction has also become popular in hot real estate markets, where sellers set offer deadlines in the hopes of drumming up competition and interest in their property. An online sealed bid can work in a variety of asset selling situations.

Live webcast auction

Finally, live webcast auctions take place in real time, usually with an in-person component allowing for bids via an online marketplace. We see these used most often in selling off transportation assets.

One Researcher's Trash…

We do what we can at REUZEit to keep assets providing value—that is, running and being an active part of the circular economy. Nevertheless, part of our cost of doing business is carrying a pool of surplus equipment. We give our partners access to this pool and have heard many great stories from folks who have put assets back into use.

Once, a research lab scientist at a company with which we'd partnered—let's call him Dr. Williams—was looking for a particular piece of equipment. He wanted to do a viability test for a new approach to a project he was working on, but he didn't have room in his budget for the expense of an automated rt-PCR machine. Even used, they'd set his lab back about $35,000. And unfortunately, there weren't any other machines available in his lab that would give him the chance to experiment with different profile arrays and methods as needed.

Dr. Williams learned that his company had recently partnered with REUZEit and that his lab had called on our pickup service to remove several stacks of incubators they'd been using for cell growth. He found out he could take equipment from our surplus pool, but assumed the pool would be full of old junk that nobody wanted. He figured it couldn't hurt to look, though, and was pleasantly surprised to find the perfect machine for his viability test sitting and waiting for him. All Dr. Williams' lab had to pay was a small redeployment fee and shipping costs from the warehouse. Not only did this save his lab tens of thousands of dollars, but he told this story in a feature shared organization-wide, encouraging reuse and resulting in the overall improvement of the organization's progress toward their sustainability goals.

Get Started: A Step-by-Step Guide to Asset Recovery

Okay. So, you're preparing to embark upon your strategic, organization-wide asset recovery journey for the first time. Don't worry. The process can be made simple with this step-by-step approach:

Step 1: Identify your surplus

Surplus isn't just the equipment that's sitting in storage. It could also be unused—or rarely used—equipment taking up valuable space in your lab or on your manufacturing floor. If you've implemented a comprehensive inventory and equipment management program, as detailed earlier in the chapter, you can probably find assets that aren't working to their fullest potential. Other indicators of surplus assets include financial reports (e.g., items that have depreciated), slow-moving inventory, and/or full warehouses or storage spaces.

Step 2: Decide what can be recovered

This chapter has focused largely on our areas of expertise (equipment in the biotech and pharmaceutical industries), but many items can be recovered to generate revenue across industries. Here are some examples, just to get your ideas flowing:

- Material handling equipment
- Vehicles and transportation assets
- Rolling stock
- Tools
- Maintenance equipment
- Testing and measurement equipment

- Production machinery and process equipment
- MRO and spare parts inventory
- IT assets

We'll get into specific examples of circular asset management in various industries in the next chapter.

Step 3: Choose an asset recovery method

As previously alluded to, there are seven key methods for asset recovery, listed here as moving from highest in value recovered to lowest:

1. **Redeploy.** Reusing equipment, or incorporating surplus assets back into the production line, keeps assets in use at their highest value. You might transition to using the assets in a more limited role, or in a different production line or location, but with redeployment they're still serving in their initial intended use.

2. **Recondition.** Refreshing or refurbishing used equipment or surplus materials to keep them performing their originally intended use, or to assign them a new purpose, is considered reconditioning. This process is often much cheaper than purchasing new and keeps materials out of landfills.

3. **Return to vendor.** As in a handful of examples we've mentioned, some vendors—from small businesses to major corporations—have buy-back policies or return programs on certain pieces of equipment. Frequently, the vendor will cover the cost of the return, and sometimes companies can get a credit toward new assets. This is also a good play from an environmental perspective, because most vendors will then put these assets (or their materials) back into their own circular systems.

4. **Resell.** When none of the above is possible, re-selling allows you to free up storage space while recapturing some of your initial investment. The downside? Management of the process, which requires time and expertise to maximize return.

5. **Reclaim.** With this strategy, teams will reclaim parts or sections of used equipment or surplus assets. This keeps some materials in use but could limit the reuse potential of the non-re-claimed (i.e., discarded) parts.

6. **Donate.** Donations get assets out of storage and can build goodwill in the community and/or give your company a good story to tell. There is potential for tax refunds, though not always, and

often the refund isn't as valuable as what your organization could get on the resale market.

7. **Remove.** Scrapping an item usually costs money rather than providing any organizational value. Ideally, you can find a way to recycle as much material as possible as opposed to sending inorganic material to a landfill.

If it helps to have a visual, this decision tree can guide your comprehensive asset recovery plan.

Asset Disposition Decision Tree

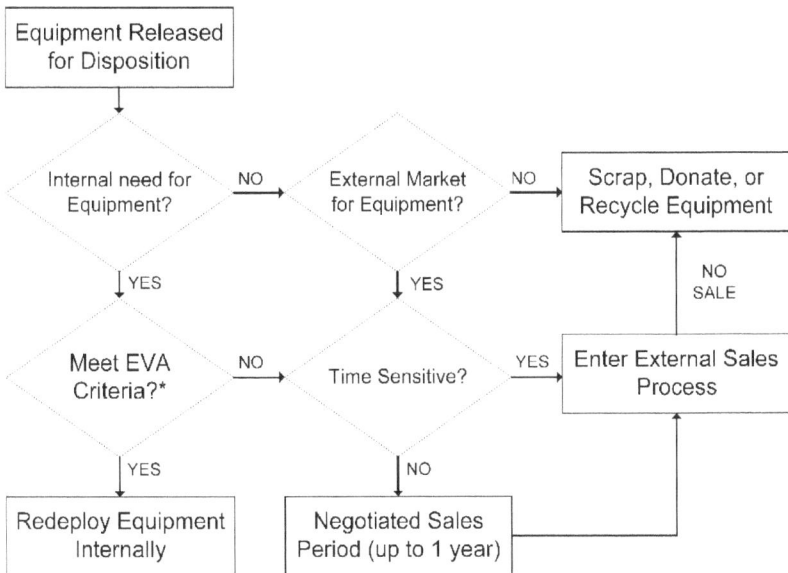

```
Equipment Released
for Disposition
        |
        v
Internal need for   NO    External Market   NO    Scrap, Donate, or
Equipment?      ------->  for Equipment?  ------>  Recycle Equipment
        |                      |                          ^
        | YES                  | YES                      | NO
        |                      |                          | SALE
        v                      v                          |
Meet EVA        NO                        YES    Enter External Sales
Criteria?*    ------->  Time Sensitive?  ------>       Process
        |                      |
        | YES                  | NO
        v                      v
Redeploy Equipment      Negotiated Sales
Internally              Period (up to 1 year)
```

* EVA: Economic Value Added. Critical to a successful program. If the transfer and shipping cost exceed the value of the equipment, internal redeployment does not make sense.

Figure 3.1

143

[The sales process and negotiated sales period mentioned in the above chart are where REUZEit gets the funding for all our other recovery and reuse operations,

making asset management financially stable.]
Having a lifecycle management plan, as well as strategies to optimize an asset's lifespan and a path guiding resource recovery, is a great setup for all these approaches. Often, in a best-case scenario, you'll find yourself using more than one of them at once for different kinds of assets.

Buckle up. In the next chapter, we'll share practical techniques and real-world examples of the circular economy/CAM across a wide range of industries.

Chapter 4

The Potential of CE and CAM Across Industries

The International Organization for Standardization (ISO) is the world's largest developer and publisher of International Standards (which are also, confusingly, called ISOs). The ISO is an independent, nongovernmental, international standard development organization[73]—for standards applying to both technical and nontechnical fields. It has bodies in 171 countries and a central body in Geneva, Switzerland. Importantly, adhering to ISO standards is voluntary; however, the standards are useful guideposts when it comes to policy, as they act as a framework that countries may use for enforced regulations or simply as references when coming up with new legislation. And individual organizations can also use ISOs as guidelines for developing their own standards, as well as communicating their commitment to certain principles with an easily understood shorthand.

In May 2024, ISO released three new international standards for a circular economy. This was a huge step that validated the importance of an all-sectors shift to circularity, as well as provided useful benchmarks—developed by CE experts from 100+ countries!—for companies, industries, and nations to measure growth and performance. The ISO's three CE standards all fall under the ISO 59000 umbrella:

ISO 59004:2024[74] Circular economy—Vocabulary, principles and guidance for implementation	*Includes the first internationally agreed upon definition of a circular economy, which it bases on six principles (Figure 4.1)*
ISO 59010:2024 [75]Circular economy—Guidance on the transition of business models and value networks	*Provides detailed actions and business-specific strategies to help leaders transition their operations to a circular model.*
ISO 59020:2024[76] Circular economy—Measuring and assessing circularity performance	*Includes detailed methodologies for selecting which indicators to use in measuring and assessing CE performance, as well as for collecting data—so that industries and organizations are, as the saying goes,*

	comparing apples to apples when looking at the outcomes of their operational shifts.

SYSTEMS THINKING
Adopting a long term strategy

SHARING VALUE
Collaborating along value chain

RESOURCE STEWARDSHIP
Close, slow and narrow
resource flows

VALUE CREATION
Maximize use of resources
efficiently

RESOURCE TRACEABILITY
Be accountable for sharing
information with stakeholders

ECOSYSTEM RESILIENCE
Contribute to the ecosystems,
regenerate

Figure 4.1

Five months later (October 2024), ISO 59014:2024[77] was published as a guidance document on the sustainability and traceability of the recovery of secondary materials. ISO 59040[78] is a product circularity data sheet that "establishes a general methodology for information exchange supporting the interoperability of circular economy related information." This should give organizations common ground when working with partners to acquire or supply products.

147

We're ecstatic to see circularity getting the attention of catalyzing bodies like the ISO. In this chapter, we'll be looking at how CE and CAM can be applied within and across all sorts of industries—beginning with one you might not expect.

CE in Nuclear Power

Assets within the nuclear power industry have a lifecycle that is much longer than those in many of the other industries we've discussed in this book. Take the extended, complex process of decommissioning nuclear facilities as an example. Increasingly, nuclear organizations are adopting ISO/TS 55010 (updated in 2024)[79], a technical specification form that provides guidelines for the alignment of financial and non-financial functions in asset management. This standard also applies to the CE, as it is crucial for an efficient and sustainable decommissioning process, emphasizing the reuse and recycling of materials to minimize waste.

Following these guidelines, the nuclear industry seeks to extend the lifecycle of assets, including the reuse of non-radioactive materials such as metals and concrete, which can constitute up to 90% of its decommissioned materials. For instance, large metal components from decommissioned reactors are often melted down and recycled into new metal products. Additionally, buildings and infrastructure can be repurposed for other industrial uses, thereby reducing the need for new construction and conserving resources. Some of this skilled decommissioning work can also keep employees engaged—if you're thinking of Homer Simpson and his job as a nuclear safety inspector, know that the decommissioning process takes 20 to 30 years,[80] and perhaps Homer could help with asset recovery until he reaches retirement.

The International Atomic Energy Agency (IAEA) supports these efforts, highlighting the benefits of CE practices in reducing decommissioning costs, minimizing environmental impact, and engaging stakeholders effectively. By integrating CE strategies and ISO guidelines, the nuclear power industry not only enhances its sustainability impact but also aligns with broader worldwide goals of reducing carbon gas emissions and promoting resource efficiency (IAEA, Vysus Group, Onepak).

We'll return to ISOs in chapter 7.

CAM in Life Sciences

We first started REUZEit as a side hustle. We pounded the pavement around "Biotech Beach," areas in Southern California that are home to lots of labs. We'd walk into reception lobbies and follow simple scripts, i.e.: "I'm a buyer with cash to buy old equipment you're not using anymore. Is there any equipment out back or in storage you'd like to get rid of?" For a year or so, this was the main way we generated income.

That all changed when we met Larry, a manager in the facilities department at Life Technologies (later acquired by ThermoFisher, expanding REUZEit into a much larger company in turn). We had first knocked on their door a year or so earlier. They never charged us for equipment, because they mostly gave us old, broken assets we couldn't resell—they were only useful for parts. But one lucky day, Larry told us about a huge warehouse (we're talking a million square feet!) that was part of an acquisition from years earlier. He said it was full of stored equipment from that initial acquisition, plus more that various departments had dumped there over the years. In truth, it was a bit of a mystery. The company didn't have an inventory on any of it.

The plan was to gut the inside of the warehouse and build out a facility with labs and offices. Demolition was slated to begin in three months, so, we had a deadline. We worked day and night to organize all the equipment into an area the size of a football field. Millions of dollars' worth of equipment. Of course, as a startup, we didn't have nearly enough cash to purchase it all, nor would we want to take on so much risk (i.e., dated equipment in unknown condition). That's why we agreed to sell everything on consignment, with the stipulation that the client had the first right to keep anything they wanted after they learned what they were in possession of.

We began cataloguing in earnest. Some of what they had was huge systems, including large tanks and equipment for processing and manufacturing specialized fluids and chemicals. Other assets were small benchtop analysers worth more than a Porsche.

Every week, we would give Larry an updated list of the equipment we'd processed, and he would share it with scientists and lab managers throughout the organization. At the time, Life Technologies has about 9,500 employees across dozens of locations. Later, as we went to the organization's sites all around the country to do the same thing—recover and catalog equipment—the list became affectionately known as "Larry's List."

The next few years were a blur. The two of us and our growing team worked to undercover millions of dollars in value that was redeployed within Life Technologies. It was a huge success for the company. And we sold all the stuff on Larry's List that went unclaimed, which paid for all our costs at REUZEit—all the travel, pickups, storage, and employees we'd hired. In fact, we took home a nice profit. Plus, we measured and reported all the sustainability metrics, like landfill avoidance and organic productivity from reusing instead of buying new.

It was from this foundation—Larry's List—that we saw what these massive, multinational companies needed: someone to handle all their valuable equipment and make sure none of it went to waste. Before us, most of these companies resorted to paying an auction company to liquidate their high value equipment (for pennies on the dollar). The low value stuff (under $5k) was usually recycled or ended up in the landfill. Far too many companies, unfortunately, still operate like this. We want them to know what all they're leaving on the table!

The Food Industry's Desperate Need for Sustainability

Human beings must eat to survive. If you think about it using the 10Rs framework, we cannot simply "refuse" food. And the system we've developed for food production has enabled the development of society, from our cities to our economies to our growing population. Lamentably, the entire food system is predicated on the take-make-waste linear model, which has taken an enormous toll on society and the environment.

Today, for every dollar spent on food, society pays two dollars in costs related to consumption and production.[81] On the consumption side, the impact is on people: micronutrient deficiency, malnutrition, and obesity. In terms of production, there are health, economic, and environmental costs. Looking at the "make" portion of the damaging triad, $5.7 trillion annually is wasted due to the way food is produced. As for "waste," the enormous amount of food waste resulting from the linear system is almost impossible to comprehend: every second, the equivalent of six garbage trucks full of edible food is either lost or wasted. Another way to look at it: we lose about one-third of all the food produced for human use, equivalent to $1 trillion. And less than 2% of food byproducts (and the resulting human waste) are recovered and put to good use. Everything connected to this agricultural waste cycle makes it the third-largest producer of carbon dioxide across all sectors. Cities are one of the biggest culprits in food waste—but at the same time, because so much of our human population is centered in and around these metropolitan areas, they can also be the catalysts in driving change.

The circular economic mindset provides an effective, attainable way to decouple food production from the wasteful industrial linear model. Within the food systems framework, CE mimics the regeneration of the natural world, eliminating waste entirely by feeding it back into the product lifecycle. To ensure all food byproducts are free from contaminants—i.e., in order for them to eventually be returned to the soil as organic fertilizer—change must start at the beginning.

Challenges in adopting the circular economy for food systems

First and foremost, there are myriad critical stakeholders in the food industry. The most obvious have models that directly profit from food production and sales: producers, brands, distributors, buyers, retailers, restaurants, and other food providers. Then there are the secondary stakeholders, who also have a role to play: waste management companies, governments, learning institutes, financial institutes, and consumers. A collaborative effort is necessary to mobilize systems-level change.

Specific challenges, or barriers to adoption, must be addressed for each of these stakeholders. For businesses, there's no avoiding that transitioning to a circular model will require an upfront investment of both money and effort. Food businesses might need to redesign products, rethink production processes and supply chains, develop new partnerships, source new materials, and invest in new equipment or technologies. Consumer behavior must also change. With intentional consumption, we reward businesses for making efforts toward circularity, thereby fuelling more change. And we'll need to adopt consumption pattern shifts to kick bad habits (e.g., overbuying and opting for out-of-season or less healthy food options). Finally, on a standards level, regulations and policies that support, rather than hinder, the adoption of circular practices need to be put into place.

Change is always an undertaking. It's one of the themes that we keep revisiting in this book for a reason. That said, it's no exaggeration to say that a collaborative effort to shift toward a healthier, regenerative, *circular* food system could truly change the world.

The potential impact of circularity in food systems

Using circular food supply chains affects people and the environment for the better. There is an economic opportunity upwards of $700 billion in reducing edible food waste and putting the nitrogen and phosphorus from food byproducts, as well as organic materials, into use.[82] An improved model would also lower health costs associated with pesticide use alone by $550 billion. To achieve these gains, there are opportunities for innovative entrepreneurs across the food value chain to tap into new, high-growth food industry sectors.

On the flipside, food insecurity and linear production methods are real threats. Despite all the food we're wasting today, 10% of the world's population continues to go hungry[83]. What's more, by 2050, our current food production processes could result in around 5 million lives lost annually due to the detrimental effects of air pollution, water contamination, foodborne diseases, and antimicrobial resistance.

Conventional farming practices have high input costs and low resilience, with threats to long-term yields due to soil degradation. But circular agriculture practices support resilience, enabling farmers to adapt to the effects of our degraded environment, thereby lowering their risk and allowing them to maintain productivity amidst shifting conditions. They also improve long-term yields, making for more productive, robust, and sustainable food systems.

Conserving nature is another vast benefit of reworking our food system. It's estimated that a regenerative model would avoid the degradation of over 37 million

acres of arable land each year and would save 450 trillion liters of fresh water.

Implementing CE in the food industry

Sharp innovators and savvy business owners are already making strides in the effort toward circular food
systems.

Packaging innovations are low-hanging fruit. Pasta
manufacturer Barilla uses its byproducts to create
packaging and paper products under the name
Cartacrusca in a partnership with paper company Fav-
ini Apeel, a California-based food technology company, created an edible, plant-based coating that can
be applied to fresh produce to get rid of single-use
packaging while still increasing shelf life. (On that note,
shelf life cannot be ignored in this rethinking process—
the primary function of food packaging is, after all, to
keep food fresh, undamaged, and safe to eat. If a packaging innovation doesn't uphold this function and results in *more* food waste due to decreased shelf life,
the involved effort toward sustainability was moot.)

In 2024, United Kingdom-based polymers developer Aquapak Polymers Ltd. conducted a study that surveyed 100 UK-based packaging experts responsible for R&D, design, and sustainability for fast-moving consumer goods brands.[84] The good news? They found that 92% planned to stop using plastic in their consumer packaging altogether within five years. Paper and paperboard were the primary replacement materials, followed by new polymers, bioplastics, and multi-materials. However, most of those surveyed (87%) wanted this switch to take place more quickly than they thought it might.

Overall, a great way to aid in the food industry's shift toward circularity is to reposition the idea of "waste" as not simply something to get rid of, but as a useful raw material. So long as they're not contaminated by pesticides or other inorganic applications, the organic by-products of food and beverage production can *always* be used in a regenerative process. As one example, grape pomace, a byproduct of the winemaking process, can be used as a significant bioenergy source; "second generation" biofuels obtained from grape pomace's biomass are easily converted into energy.[85] There are a few hurdles to jump through to establish regular production, but doing so would avoid the extraction of raw materials for energy use altogether. The opportunity is ripe—pun intended—for widescale implementation.

When it comes to preventing food waste at the source, retailers can institute waste reduction practices such as discounting soon-to-expire items, carefully tracking demand for products to try to match supply as closely as possible, and/or donating excess inventory. Trader Joe's has a Neighborhood Shares program[86] that donates "100% of products that go unsold but remain fit to be enjoyed by communities seeking food assistance." Each neighborhood store partners with a local nonprofit to provide them with food nearing its expiration date. In 2023, the company donated more than $469 million dollars' worth of products, or over 104 million pounds of food. Cities can also play a role. At the end of a food product's lifecycle, effective collection systems for organic waste (e.g., for composting), and pure waste streams that make use of processes like anaerobic digestion, extract as much value from each item as possible.

Despite such impressive opportunities, and although implementing regenerative practices often pays for itself in the long-term (with previously mentioned benefits like resilience and increased yield), the upfront costs can be hard to swallow. Keep in mind that consumers are willing to pay more for foods/products from regenerative farms. And in a crowded marketplace, highlighting certifications or nutritious ingredients in product labeling can be the reason a consumer purchases a product in the first place. In 2016, the Pew found[87] that 73% of Americans reported purchasing locally grown food in the last 30 days, 71% decided to buy based on ingredients and/or nutrition label information, 68% bought organic food, and 44% bought food labeled GMO-free. Highlighting any of these properties can be a great way to both draw attention to your products and increase their value in the eye of the consumer, making people more willing to pay a premium. Along these lines, educating consumers and raising their awareness of the benefits of these foods is a worthwhile part of a regenerative CPG food brand's marketing strategy.

As another selling point for industry circularity, 2014 research by an Australian scholar found that the industrialized food production system will face challenges from both pollution and peak oil (the hypothetical point in time when the total production of oil reaches its maximum rate). Community gardens are a pragmatic safeguard against these concerns, especially in urban areas—a beautiful solution that can also put unused land to use, produce healthy local food, and engage neighbors is community gardens. While they may seem quaint, or romantic, or even frivolous, at their best, community gardens can also address food insecurity and guard against possible food shortages.

A roundup of outreach efforts by land-grant universities in the US[88] confirms the impact of community gardens and related efforts on increased food security and community wellbeing. As one case study[89], an urban garden in Georgia was established to support a resource center that provided meals for the local unhoused population. The garden supplied over 3,000 pounds of fruits and vegetables for the center's kitchen in one growing season, including peas, beans, okra, peppers, tomatoes, corn, squash, strawberries, watermelons, and sweet potatoes. The estimated 3-month food cost savings across over 23,000 residential meals was $10,000—and the menu got much healthier.

CE in Education

Now, let's consider another industry with plenty of opportunity for circular innovation—education. Higher Education Institutes (HEIs) are poised to play a huge role in the transition from a linear to a circular economy via their campus management efforts and deliberate educational models, some of the compounded outcomes being the influence a CE mindset can have on the coming generations of citizens and leaders, as well as private and public stakeholders who may collaborate on these efforts. Plus, many look to HEIs as a hub of innovation from which to learn.

Many HEIs have formally committed to adopting a sustainable campus; however, most fall short of wholesale rejection of linear processes in favor of circular initiatives. Sometimes, CE is only considered by HEIs with regards to waste management practices, whereas opportunities exist across the entire economic model and resource management strategy. Campuses can be thought of as small-scale cities, with CE strategies being applied across areas including teaching, research, campus management, student-led projects, influence, and leadership.

Notably, HEIs are economic drivers because of their purchasing power. By building a closed-loop supply chain and using circular strategies for both procurement and waste management, HEIs will influence their suppliers and partners to develop their *own* CE strategies, causing a ripple effect outward. A couple of major things for HEI leaders to consider are seasonal variation in flows of people and campus activities, as well as the wide-ranging nature of said activities and resulting miscellaneous waste composition.

For HEIs to holistically plan and maximize circular actions, leaders must adopt a systems-oriented way of thinking. According to the article "Higher education institutions as a microcosm of the circular economy" in *The Journal of Cleaner Production*, "The use of systemic analysis is necessary to provide a multidimensional and dynamic perspective that considers the different scale levels and the complex social, economic, technical and political contexts related to the implementation of CE in HEIs."[90]

Another way to approach a CE mindset in HEIs is to consider asset management. Recall our focus on circular asset management (CAM) in chapter 3—and our story about donating microscopes to schools? Learning institutions manage all sorts of assets as part of their regular operations, and HEIs can reduce their resource consumption/ emissions via the shared or joint use of assets—either with other schools or other community groups. Specifically, schools have spaces that frequently go unused outside of school hours. By opening sporting facilities, performance venues, playgrounds, and classrooms up to other members of the community, resource use is better optimized. (In certain instances, it may make sense for school-owned property and/or assets to be leased by third parties.) Furthermore, schools may find their end-of-life assets, such as older computers or underutilized school buses, in demand on the resale market; in turn, they may put any new, resulting income toward purchasing newer equipment/assets or providing onsite ancillary services (e.g., after-school and summer care programs).

The education system is also the perfect starting point for teaching future generations about the importance of circular actions. When school systems and HEIs move to adopt circular models, not only do they benefit the environment and their bottom line as organizations;

they also act as case studies of the model for their students, faculty, and staff—as well as the family and friends of all of these folks, their suppliers and partners, their alumni, board members, visitors, and beyond.

Education for Sustainable Development (ESD) is a UNESCO initiative to incorporate the United Nations Sustainable Development Goals in education. Each year, UNESCO awards a prize to three organizations making notable strides. One of the 2023 winners was Long Way Home in Guatemala, for their Hero School project, the organization's flagship PreK–11th grade school located in San Juan Comalapa, Guatemala. The 20-building campus was built, in part, from 500 tons of rubbish and 25,000 discarded tires. As stated by UNESCO[91], the school:

> "Aims to provide affordable and high-quality education to marginalized communities while integrating sustainable values into lesson plans and curriculum. Such sustainable values include pollution mitigation, improved access to quality education and clean water, food security, gender equality, and improved human rights.

Students actively contribute to the community's well-being by constructing essential living conditions and structures such as smoke-reducing stoves, drinking water cisterns, safe waste disposal latrines, landslide-preventing tire retaining walls, and earthquake-resistant safe housing.

The project garnered the jury's recognition for its transformative impact and educational approach. Their efforts in empowering students, training teachers in sustainable values, and addressing economic and environmental challenges in a holistic way were commended, particularly in providing marginalized communities with knowledge and resources for sustainable self-reliance."

Schools aren't the only educational spaces that can be hubs of knowledge around and models for CE principles. The city of Chefchaouen, Morocco, has also been recognized by UNESCO[92] for its ESD implementation, specifically for its Energy Info Centre (EIC). Launched in 2016, the EIC is a nonformal education center that aims to advise citizens on energy consumption. According to UNESCO's report "ESD Implementation in Learning Cities," the center's "priority areas include the reduction of electricity and water consumption; the use of solar energy; the improvement of air quality in indoor spaces; the mitigation of and adaptation to environmental degradation; the safe use of gas equipment; and more efficient management of household and business waste." The center has a test kitchen stocked with energy efficient and non-energy efficient appliances for comparison, as well as educational resources and energy usage measuring equipment that it loans out to residents. The program faces challenges, but its goal is to change collective norms and standards with regards to citizens' energy consumption so that the municipality can localize and implement its Sustainable Development Goals by 2030.

The Pressing Need for Circularity in the Retail Industry

Perhaps more than many other industries, the retail industry has long been virtually synonymous with the take-make-waste linear model. In chapter 1, we mentioned how textile production uses about 4% of our total freshwater supply and that shifts to circularity within the textiles industry amount to a $500 billion economic opportunity. At this point, shifting to CE is moving from a nice, potentially profitable effort to an industry imperative. The linear model is showing its great weaknesses; factors like resource scarcity and associated rising costs, supply chain breakdowns, tariffs, regulatory restrictions, consumer demand, and brand competition are converging to put more pressure than ever before on retail organizations to change, and quickly.

Deloitte's 2023 CxO (C-suite level business leader) Sustainability Report[93] found that 73% of CXOs had increased investments in sustainability over the course of the year—with more adopting increasingly complex strategies to better move the needle. (Some of these "harder to implement" actions might represent larger barriers due to time, effort, and investment required, but many of them have the greatest potential for long-term, sustainable impact.) The following chart, developed from 2023 and 2024 data pulled from these Deloitte reports shows how retailers are designing their strategies.

Deloitte's 2023 CXO (C-suite level business leader) Sustainability Report found that 73% of CXOs had increased investments in sustainability over the course of the year. The following chart shows how retailers are designing their strategies.

TOP ACTIONS TAKEN

Which of following actions/adaptations has your company already undertaken as part of its the sustainability efforts? (select all that apply)

2023		2024
65%	Using more sustainable materials (e.g., recycled materials, lower emitting products, less plastic material, 'circular' products)	51%
62%	Increasing the efficiency of energy use (e.g., energy efficiency in buildings)	50%
53%	Using energy-efficient or environmentally-friendly machinery, technologies, and equipment	49%
50%	Developing new climate-friendly products or services	49%
48%	Training employees on carbon footprint actions and impact	48%

HARDER TO IMPLEMENT, NEEDLE- MOVING ACTIONS*

2023		2024
50%	Developing new environmentally-friendly productsor services	48%
48%	Tying senior leaders' compensation to environmental sustainability performance	47%
43%	Requiring suppliers and business partners to meet specific sustainability criteria	46%
38%	Updating/relocating facilities to make them more resistant to pollution impacts	44%
27%	Incorporating environmental considerations into lobbying/ political donations	43%

*As defined by Deloitte's analysis

2023 Deloitte CXO Sustainability Report

In particular, resale is an industry strategy that's well aligned to the consumer mindset. The market for secondhand clothes has long existed through channels like thrift stores, vintage clothing boutiques, garage sales, clothing swaps, and bins full of hand-me-downs passed among friends and family. But as Millennial and Gen Z consumers—who are increasingly preferring sustainable products and unable to afford regularly purchasing firsthand products due to soaring costs—take over a larger chunk of buying power in every market, brands are taking notice in more strategic ways. Deloitte notes: "We expect future retail models, especially those used by large companies, to combine mainline, outlet, and resale models...it is only logical that brands would want to control their own resale channels. This allows companies' profits to grow and reduces reliance on production channels."[94]

This cascades nicely into the next effort retail leaders are making in CE: sustainable supply chains. The retail supply chain contributes to 25% of current worldwide carbon emissions, and the world is suffering as a result. A more resilient supply chain built around circularity is a more complicated endeavor than, say, making a straightforward materials swap, but the effort can truly accelerate progress toward a company's sustainability goals. Moreover, as we've seen in other industries, improved traceability and increased efficiencies from eliminating waste from the value chain can result in both cost savings and improved customer satisfaction and retention.

Newer technology like blockchain and artificial intelligence (AI) can enable progress on sustainability efforts in the retail industry. Blockchain increases traceability and transparency across the supply chain, highlighting opportunities to improve efficiency. We'll dive into AI further in chapter 6, but it's worth mentioning here, as well, that AI can also improve traceability. Furthermore, AI can aid in designing circular products and increase the accuracy of forecasting—all of which can reduce waste throughout product lifecycles.

The good news is that retailers are making strides towards circular actions. H&M, for example, is using circular fashion to drive foot traffic in its brick-and-mortar locations by offering recycling points in stores.[95] Since launching their garment collection program in 2013, they've received nearly 190,000 tons of textiles. The program allows shoppers to return old clothing of any brand for reuse (selling secondhand), repurposing (into other textile products like cleaning cloths), or recycling (usually downcycling the textile fibers into things like insulation material).

In 2020, UK supermarket giant ASDA introduced a refill zone in one of their stores; as of late 2024, it has them in four. The concept is loose-format shopping: customers can use their own containers to buy over 75 brands of unpackaged products, encouraged by a "refill price promise" guaranteeing loose-format goods will always be cheaper than prepackaged, and via coupons and rewards for purchasing from the refill zone. This initiative is part of Asda's goal to remove 3 billion pieces of single-use plastic from their business by 2025.[96]

Lastly, in the retail industry electronic waste is particularly damaging—because it often contains materials that are toxic to human health and the environment. And unfortunately, the volume of e-waste has nearly doubled since 2010, resulting in 62 million metric generated annually, worldwide.[97] Apple is trying to turn the tide, with a goal of zero net emissions by 2030.[98] Their efforts include removing plastic from their packaging, with over 97% of their packaging being fiber-based as of late 2024; making packaging more compact; and recycling and reusing as many critical materials as possible, including nearly 40,000 metric tons recycled in 2023.

To be clear, it's not just high-tech and household brand names that are making notable steps toward circularity. Some of the oldest industries you can imagine are also capitalizing on CE opportunities.

CE-SAM in Mining

Some circular principles are already in place at many mine sites, including those that reduce waste and conserve water, because the impact on the bottom line has been proven out. However, the best value will come when every lever, from production to recycling, has been pulled. To put it briefly, a holistic shift to CE in the mining industry—one that functions across the entire value chain—requires collaboration among stakeholders including policymakers, designers, and industry titans.

Researchers have found that[99], especially in developing economies (like, for example, Namibia), the mining sector's transition from linear to circular production systems is slow, and companies are heavily reliant on government policy and tax incentives for adoption. So, which policies stand to make the biggest difference in accelerating closed-loop systems in mining? Those that consider coherence between sustainability goals and society's material needs. Those that use the right metrics to accurately measure performance. Those that place appropriate incentives on efforts to remove barriers to circularity. And those that encourage advanced models—which might be harder to undertake solely based on market-driven value decisions, but will drive the biggest impact.

Depending on your relationship to mining and the connotations the industry carries for you, you might hesitate to recognize the ways in which the industry is key to shifting toward clean energy. But copper is widely seen as a linchpin of this transition, with the annual worldwide demand for refined copper expected to double by 2050.[100] As a material, copper is durable and versatile, with a lifespan of up to 50 years in some applications. Renewable energy sources like wind and solar, storage batteries, and electric vehicles all rely heavily on copper. Furthermore, about one-third of current demand is met by *recycled* copper.[101] The goal, then, is to enhance efficiency at a large scale—to keep the material in use at its highest value for as long as possible.

In addition to a clear focus on optimizing the extraction, processing, and lifecycle of copper, a handful of other circular initiatives in the mining industry are worth highlighting. Most center on reducing waste and pollution at mine sites and enabling regenerative efforts. The International Copper Association[102] buckets current initiatives into three groups: waste to value, waste reduction, and land generation. Waste-to-value efforts include recovering steel from discarded haulage truck tires and recycling it into grinding balls used at the mine. During one such pilot program, Antofagasta's Los Pelambres mine aimed to repurpose 97 tonnes of steel from 156 tires. Other mining groups minimize water waste, establishing ways to reuse water by desalinating water supplies. Finally, Teck's Highland Valley copper mine works toward regeneration of previously mined areas by replanting native trees and reintroducing wildlife to restore biodiversity.

The International Copper Association puts it well: "As the circular economy gains momentum, mining will play an indispensable role in shaping a more circular world and a sustainable future for all."

One of the most common CE applications in the mining industry involves reprocessing tailings materials (the waste left over from the process of extracting metals and minerals from ore) to extract any resources that may still have use. But new perspectives can enable business owners to understand just how much more opportunity there may be in their own current models. For example, reusing or donating used equipment, as we explored in the previous chapter, could have positive environmental and revenue benefits for mining companies. A simple mindset can make a huge difference—i.e., if owners look at themselves as *borrowers* of the land and think about what value they can add (rather than just extract) to the areas they process. And taking a long-term view, focusing on regeneration and facilitation can bring broader benefits to the mine's greater region. *CIM Magazine*[103] quotes Alan Young, one of the directors at Materials Efficiency Research Group:

> "...frankly, it's [mostly] been a harm reduction model. How do you do less harm? How do you

minimize the impact or mitigate stuff? That's all good and important, but there's a powerful sense that this is more of a transformation piece. How do you move to net benefit? How do you really move into closed loops? And how can the industry rise to this moment where we need a lot more metals, and we can't justify a conventional industrial model for a variety of reasons?"

In summary, there are a handful of sustainable approaches that many companies will find worth exploring: use of renewable energy resources, water management, responsible waste management, and environmental rehabilitation and restoration. As is a theme throughout this book, each of these offer environmental and financial benefits once implemented, and many of them can be launched in tandem. Specific ways to approach circular actions in the mining industry will be highly dependent on a business' current infrastructure, product focus, and operational models.

Aviation CAM Solutions

While circular economy models aren't as widespread in the aviation industry, this mostly signifies to us that the industry is full of potential. As an early indicator that the industry's mindset is currently shifting, the International Civil Aviation Organization's (ICAO) mentions CE on their website,[104] saying that CE concepts could "provide valuable environmental, social and economic opportunities." (Established in 1944, ICAO is a United Nations agency helping "193 countries to cooperate together and share their skies to their mutual benefit.")

Innovators beginning to disrupt traditional flight models include the company Joby Aviation. Launched in 2009, Joby has developed a pioneering electric aircraft powered by six electric motors that takes off and lands vertically. These aircrafts generate zero operating emissions and require far less infrastructure than airports and runways. Joby has signed deals with Toyota and Delta with the goal of providing aerial ridesharing services.

On a more tactical, commercial airline level, we can look at the entire flying experience, including in-cabin and in the airport, to find ways to incorporate circular measures. In 2023, per the International Air Transport Association, airlines generated 3.6 million tonnes of cabin waste.[105] While many airlines are interested in reducing, reusing, and recycling efforts to improve upon these numbers, both the nature of air travel and strict regulations regarding the handling of waste present challenges. Plastic wrapping is often used because of rigid food safety, hygiene, freshness, and weight requirements. Moreover, many countries require that airlines treat catering waste as high-risk, as a precaution against international transfer of diseases, so they must either incinerate it or bury it deep in landfills. Advocates argue that smart, tailored regulations could maintain health controls while allowing for more recycling.

Even with these hurdles, a handful of innovative sustainability strategies are already gaining traction across the industry. To start, airlines can use data to track food and drink consumption over time, so they can stock flights in a way that avoids overstocking and curbs waste. Next, some airlines are asking travelers to order meals before their flight, a pay-as-you-go approach that reduces the amount of untouched food that gets tossed. By reusing still-sealed unused food items on future flights, Air New Zealand has diverted more than 1,500 tonnes of in-flight waste from landfills. Other airlines are finding alternatives to plastic for packaging and cutlery: Emirates uses plastic meal trays and bowls, but with a closed loop program that involves washing and grounding them down so the material can be used for other products.

Recycling approaches outside of *in-flight* waste show promise, as well. Airports are ecosystems of their own and many are implementing promising sustainable waste management systems. At London Stansted Airport, over 150 tonnes of coffee waste is generated each year; in 2019, they began converting this waste into coffee logs that are used as biofuel, in wood burners and multi-fuel stoves, resulting in a net CO_2 emissions savings of 80%. Vancouver International Airport created a target of diverting half the waste produced at the terminal from landfills by 2020—and they achieved their goal early, in 2016, thanks in large part to a successful organic waste recycling program. Finally, Portland International Jetport captures extra aircraft de-icing fluid after it's sprayed on planes so effectively that they became the first airport in the United States to use fluid that is 100% recycled. Looking elsewhere in the industry, material from used employee uniforms has been repurposed to create first aid pouches (Korean Air); laptop bags, travel kits, home insulation, and punch bag filling (Delta); and tables and benches (Finavia). Easyjet starts farther up the value chain with cabin crew and pilot uniforms made from recycled plastic bottles.

Finally, aircrafts themselves are massive pieces of machinery, and considering their end-of-service treatment is where circular asset management comes into play. After 20–25 years of service, an aircraft usually reaches the end of its useful life, at which point 85 to 90% of it (by weight) can be recycled. Manufacturers are thus charged with expanding circularity at the beginning of an aircraft's life, by designing for durability, efficiency, and maximized end-of-life opportunities from the get-go.

Justin's Experience in the Brewing Industry

Justin can often trick people in the two-truths-and-a-lie icebreaker. "I can put official letters after my name," he says. "D.Brew." And it's true. He has a diploma from the Institute of Brewing and Distilling, the industry's foremost group of certifications. It's a big deal, though he barely used it as he expected. . (He had one internship as a professional brewer, and didn't like dragging hoses in the cellar at three-thirty in the morning. Plus, he was taking the "night-shift" for his newborn son so brewing wasn't going to workout. The science and the systems, love it. But not making beer recipes all night long.)

Still, Justin's brewing credentials intersected with our REUZEit work when we heard about a small brewery in Washington state that was dealing with a bottleneck (pun fully intended) in their production process. They brewed amazing, complex Belgian-style beers with huge body. But their blonde ale took a full five days in the brite tanks (used for carbonating beer, or storing it before packaging) to settle, or clarify, into its bright, blonde namesake color.

Well, another client of ours working in cancer research just so happened to have a superfluous contiguous centrifuge they'd used for clarifying milk proteins. We brought this to the brewery, along with a bowl that was suitable for filtering beer. The brewers were suddenly able to clarify the beer in hours, rather than days, quadrupling their blonde ale production with no changes to recipe or resulting look, aroma, flavor, or mouthfeel. From cancer research to beer brewery—the innovation

CE makes possible still sometimes amazes even us. Imagine this at scale, with best practices and human assisted AI, any equipment profile, and reuser profile to match an ideal new home for equipment to be used again and again.

Get Started: Making Connections Across Industries

Whether or not you're in a position to take apart an airplane for surplus asset recovery or source leftover centrifuges, you might be surprised how your professional and educational experiences up to now have prepared you to be a creative thinker when it comes to circular approaches. One of the most exciting parts of being at the forefront of the CE revolution is witnessing all the innovation we're seeing across industries. While you're thinking through and forming a CE business model of your own, for your company or within your organization, there are plenty of eco-friendly business solutions that can be put into place right now. Start by doing a little bit of research to see which other industries use the equipment you also use every day—you might be surprised at what you find!

Chapter 5

Eco-Friendly Business Solutions

U p to this point, we've focused on what CE is and how to put it into practice. Now, it's time to look at the other pivotal side of this work: how these initiatives we're talking about translate into your business's sustainability agenda. Transitioning to a circular economy takes time, but there are tons of actionable ways to join the fight against limited resources, overproduction, pollution, and waste…starting today.

Let's talk about an industry we haven't mentioned yet, one where progress is promising but much more work remains to be done: the maritime industry. Did you know that maritime transport is responsible for 80–90% of the world's trade volume?[106] Because of its scale and *sector-effective* impact, widespread application of even the smallest eco-friendly, circular solutions within this industry can have an outsized positive effect on the environment. For example, researchers found that a remanufactured maritime engine costs just more than half what a new one costs, while reducing material consumption by up to 30% for an engine block/crankshaft and up to 98% for a turbocharger. The win for both the ship owner and the environment are clear.

But at the same time, unfortunately, the maritime industry does not have an efficient reverse supply chain system in place, and there is little implementation of CE principles to date. An organization like REUZEit could help ensure the recovery and remanufacture of high-value assets at the end of their useful life, connecting the top shipyards and boating companies across the world to keep assets in use. A robust database that engages vertical AI and manages assets across their lifecycle would allow for holistic resource management across the industry.

Motivation for Going Green

As we've argued thus far, "going green" in your business model is good for both your profitability and the environment. But we can expand the thinking even beyond that.

One roadmap you can use to guide your business down the path of sustainability is the triple bottom line. In addition to *profits*, the triple bottom line adds two more stakeholders to the equation: *people* and *our environment*. (Importantly, this does *not* imply focusing on sustainability at the expense of profit. Profits matter. In a capitalist economy, strong financial performance is essential for continued viability. But with this framework, profits are not the sole focus, and efforts toward business growth cannot harm people or the environment.)

The second stakeholder, *people*, refers to a business's societal impact (i.e., its commitment to people). Essentially, a business should not just be committed to its *shareholders* and providing them with value; it should also work toward creating value for all *stakeholders*, including customers, employees, and members of the community. This can be accomplished by paying employees a fair wage and providing them with benefits and work-life balance. Nonprofit partnerships are another way companies demonstrate their commitment to society at large.

The third benefactor to triple bottom line is our environment—in which businesses commit to making a positive impact on the world—have the strongest connection to principles of circularity. As a Harvard

Business School report puts it,[107] "While businesses have historically been the greatest contributors of pollution, they also hold the keys to driving positive change."

As many of the case studies we've explored have already illustrated, enacting a triple bottom line strategy usually brings about financial benefits in due course. Circular asset management, as discussed in chapter 4, helps organizations cut costs while minimizing resource extraction and reducing carbon emissions. But let's take things one step further, considering how sustainability and financial performance are explicitly correlated. Environmental, social, and governance (ESG) metrics are the most widely used third-party measurement of a company's triple bottom line procedures and adherence. Once again, Harvard Business School states:[108] "… evidence has increasingly shown that firms with promising ESG metrics tend to produce superior financial returns. As a result, more investors have begun focusing on ESG metrics when making investment decisions."

On the consumer demand and loyalty side, a 2022 IBM study found that half of consumers say they're willing to pay a premium for sustainability, and purpose-driven consumers—those who choose products and brands based on value alignment—are the largest market segment (44%).[109] Simon and Kucher's 2021 Global Sustainability Study found that 63% of consumers had recently made modest to significant shifts towards being more sustainable, and when minor purchasing changes were included, this number reached 85%.[110]

In a word, the market is moving in the direction of sustainability no matter what. Strategically shifting to pursue the triple bottom line is one way to get your organization where it needs to be *before* you lose market share to competitors who are connecting better with these growing numbers of sustainability-minded consumers. It's time to get onboard! While concerns about increasing operational costs are, of course, valid—change often rings alarm bells, and rising or even unknown costs are usually among the first presumed barriers to any business transition—such barriers might indeed be opportunities in disguise. We see time and time again how operations are usually *optimized* with sustainable measures (i.e., decreasing energy use/resource consumption and increasing efficiency)—so costs usually go down and investments pay for themselves.

Office Supplies

One of the simplest efforts just about anyone in any organization can make toward sustainability involves the big umbrella of office supplies. Beyond personal steps taken, helping establish an organizational green committee or setting a meeting with your office manager to discuss ways to be more eco-friendly in your day-to-day office environment are two great ways to start. In this section, we've rounded up some eco-friendly office supply ideas that are widely adoptable.

First, go paperless. The average office worker uses 10,000 sheets of paper each year,[111] and nearly 3.7 million tons (or more than 700 trillion sheets) are used annually in the US. And even though this paper is highly recyclable, the US EPA estimates that paper and paperboard account for almost 40% of the nation's garbage. While switching to digital notebooks and using cloud storage can help, for instances when paper is truly needed or strongly preferred, buy post-consumer recycled and/or chlorine-free brands, and be sure to always use both sides of a single sheet.

Other everyday office items can be sustainably sourced as well—e.g., recycled ink and pens, and eco-friendly cleaning products. Toner cartridges and ribbons for printers can often be refilled and reused. Non-toxic, biodegradable cleaners are better for the health of both your employees and waste streams. When it comes to stocking an office break room, eliminate single-use cups, filters, and utensils, source reusable or biodegradable alternatives, and encourage employees to choose reusable containers for their coffee, tea, and water.

Another simple idea that has been proven to have great success is eliminating individual waste bins in favor of communal stations with separate recycling, compost, and landfill bins. Etsy tried this way back in 2013 in their Brooklyn headquarters.[112] They found that their waste dropped 18% while their recycling and composting rates rose by 20% and a whopping 300%, respectively. We suggest using compostable trash bags and seeing if you can donate your compost to a local community garden.

Beyond physical waste, an easy update like swapping out your current bathroom faucets for water conservation versions can make a dent in the water your employees use, therefore shaving down your water bill. And beyond providing aesthetic benefits, adding plants throughout the office improves indoor air quality because they can absorb pollutants, weed out toxins, and restore air's water vapor levels. This, in turn, can make employees happier, healthier, and more productive. Aloe vera, snake plants, money plants, Ficus trees, and peace lilies are some of the best air purification options.

Lastly, opting for sustainable office furniture or reusing old office furniture and equipment are both important components of a green office strategy. High-quality, sustainable furniture may be more expensive upfront, but it often has a longer lifespan, saving you money in the long run. Steelcase, Herman Miller, and Knoll all have sustainable office furniture lines. In-house, furniture or equipment that is no longer used can be given to employees—helping them set up a home office for remote work or check-ins—or repurposed somewhere else in the organization, and unused office desks might find a place in the breakroom, for example. Donating gently used furniture or equipment is another great way to keep items in use. Asset Network for Education

Worldwide (ANEW) is a nonprofit that was founded in 2005 to extend the circularity of surplus office items, and their Surplus Stewardship program matches corporations with charities, nonprofits, public agencies, and more that can keep their furniture and other equipment in use. ANEW keeps around 3 million pounds of materials out of the landfill on average each year.[113] More than 2,000 organizations in 20 countries have been on the receiving end of these goods.

Equipment

Equipment is at the core of our expertise, and we're full of ideas on how to make better, more sustainable use of your office equipment. Here are a few highlights.

Begin by identifying any equipment that is no longer in use or not needed within your organization. And don't let the idea of getting started feel daunting—you don't need a robust database to develop a reuse program. Start small and simple. Make a list of all equipment and make it available in a central, easy-to-find place— somewhere that can be shared throughout the company (e.g., on the intranet, a Google Sheet, or in a physical checklist or binder). Label each piece of equipment with the contact information of its last owner(s)/operator(s). Make sure to include a box next to each item for designating any necessary software, manuals, wires, cords, service logs, or other accessories that are needed to operate the machine or system. Another straightforward getting-started technique? A donation system. Create a donation area and let all employees know where it is, so they can take the initiative to drop off their own unused supplies, furniture, and equipment. Warehouses work well for this, but early on, a storage closet with extra space or even an unused area beneath stairs can also suffice. If you don't have a reuse program, you can at least partner with a local nonprofit for regular pickup (i.e., weekly or monthly) of these items, hanging signs or sending out messages to let staff know when the next planned pickup will occur. [We've seen this practice used most frequently with office supplies and furniture, but it also works with lab consumables. Some of our clients collect unused pipette tips or personal protective equipment (PPE) that is picked up monthly by a local university.]

Let's consider one example. We got the above dedicated donation area idea from one of our life sciences clients. When we got our first official REUZEit truck, one of the facility managers we worked with asked if we could stop by on the last Friday of every month to pick up equipment they no longer needed. He managed a large facility with several labs, so he communicated with all the scientists and lab technicians to put broken and unused equipment into a dedicated space in their warehouse, rather than dumping it in their recycling bins. And, as requested, we picked it all up, once a month.

This common-sense innovation resulted in millions of dollars in capital recovery and thousands of tons of landfill avoidance. Plus, because we were able to sell much of the equipment we picked up, profits more than paid for the pickups—meaning the program had a net zero cost. This program is still highly successful to this day. And now, we're driving all over with a small fleet of trucks, collecting surplus equipment from multiple clients. When you scale this work—monthly equipment pickups across multiple sites, industries, economies, nations—you change the world. And it all tracks back to that facility manager and our first truck.

A couple slightly more advanced equipment solutions involve monitoring systems. We talked about real-time equipment monitoring systems in chapter 3 and want to bring them up again here. An on/off monitor on high-value equipment can help pinpoint underutilized assets. (If a piece of equipment is turned off for a long period of time, the monitor can alert the equipment owner, prompting them to consider making the equipment available for the reuse program or to see if there's any interest in sharing it across departments or functions.) These monitors can also identify equipment fatigue and predict failure, which can be highly useful for operations and maintenance—especially if preventative measures can save the equipment before an expensive loss occurs. An added benefit is the digital history of a machine's performance these tools provide; refurbishment is more efficient and effective when techs have this data and can directly see which components are worn and may need replacing.

Managing Energy Usage

Updates to office equipment and supplies, as mentioned above, can help with energy usage in your office. The National Renewable Energy Laboratory, part of the US Department of Energy, put out a report in 2022 called *Operational Emissions Accounting for Commercial Buildings*[114] explaining that buildings are the largest consumers of energy and one of the largest sources of greenhouse gas (GHG) emissions. Specifically, about 70% of electricity use, 40% of total primary energy consumption, and 30% of operational GHG emissions in the US can be attributed to buildings. Establishing green energy practices will not only put a dent in these numbers—you'll also notice a difference in your monthly bills, and you may be eligible for tax credits, deductions, and incentives to help reduce your operational costs even further.

Here are a handful of ideas to help you cut down on your company's energy usage:

- **Increase natural light.** Designing a workspace with plenty of natural light allows you to turn off artificial lights for much of the day, saving on costs and boosting employees' mood and productivity. 2018 research from Cornell University[115] found that optimizing natural light in office environments decreased worker headaches by 63%, drowsiness by 56%, and eyestrain by

51%—significant health and wellness gains sure to improve productivity and overall morale. (We're certainly in better moods and feeling a lot more creative when our heads don't hurt, and we aren't tired!) In fact, another 2018 poll of 1,614 North American employees found that "access to natural light and views of the outdoors are the number one attribute of the workplace environment, outranking stalwarts like onsite cafeterias, fitness centers, and premium perks including on-site childcare."[116]

- **Switch to energy-saving lightbulbs.** The US Department of Energy maintains that energy-efficient lighting—like LED, CFL, or T5 fluorescent bulbs—uses at least 75% less energy and lasts up to 25 times longer than incandescent options.[117] While cost savings depend on your usage, some simple math will help you see the impact on your operations. (And, buying bulbs that last longer means less waste in landfills.)

- **Turn off everything at night and unplug.** Motion-activated light switches are one way you can cut down on your energy use automatically. They're especially well-suited for conference rooms and other shared spaces over which people are less inclined to feel ownership. Next, unplugging is more important than you might think. As reported by the US EPA, the total electricity consumed by idle electronics—i.e., "vampire power"—equals the annual output of 12 power plants.

- **Use smart plugs and/or smart outlets.** Building on the above, if your employees can't be bothered to unplug things, or switch off a power strip, smart plugs/sockets are a good solution. They're hardwired and simple to install—just plug them into your standard outlet and they turn off lights, fans, and appliances automatically. (They can also be controlled via an app on your phone. You can even set schedules!) Smart *outlets* are more inconspicuous, but they are more difficult to install and often cost more.

- **Make the switch to renewable energy.** Reducing your reliance on fossil fuels in favor of renewable energy sources is both good for the environment and can reduce costs substantially over time. The cost of electricity from solar power is becoming cheaper and cheaper,[118] meaning that installing solar panels pays for itself more quickly than ever before. Added benefits include increasing your property value, access to tax breaks, and protection against power outages via reduced reliance on an electrical grid. Plus, if a customer can see solar panels on your building, or you use them to tell a story about your company's focus on sustainability, your brand may get a boost, too. Building on this idea, adding a solar-panel cover to an existing open parking structure is a great two-for-one strategy, protecting employees and their vehicles from the elements and providing your building with a clean energy source in one fell swoop.

- **Consider efficient measures for heating and cooling.** Monitor heating and air conditioning use, keep doors and windows closed when necessary, and make sure office spaces have good insulation, especially around windows. Consider using a smart thermostat that allows you to schedule temperature controls based on when your office or facility is or isn't occupied.

- **Buy/rent/use energy-efficient equipment.** Invest in assets with an eye toward (a) circularity, and (b) having the smallest possible footprint while in use.

Reducing Employee CO2 Emissions

Now is a good time to touch on the different types of gas emissions. When a company begins down the path toward net zero, they must establish a GHG emissions baseline, meaning they need to measure and assess their current emissions state. The Greenhouse Gas Protocol, the world's most widely used GHG accounting standard, has created three buckets or "scopes" under which a company may categorize its emissions. As they explain,[119] "Developing a full emissions inventory—incorporating Scope 1, Scope 2 and Scope 3 emissions—enables companies to understand their full value chain emissions and focus their efforts on the greatest reduction opportunities." What follows is an explanation of each of the scopes:

- **Scope 1 emissions:** Those from sources an organization owns or controls directly. (E.g., If a company owns a fleet of (non-electric) vehicles, the emissions from the fuel these vehicles use.)
- **Scope 2 emissions:** Those from a source where the energy a company uses is produced (e.g., emissions caused by the electricity that powers the company's buildings).

- **Scope 3 emissions:** Any of those not covered in scope 1 or 2 that are still created by the company's value chain.

Scope 3 emissions account for the highest proportion of total emissions by far, and they are also the hardest to measure and reduce. For example, one of the emissions categories considered Scope 3 is those coming from employee commutes. And while you cannot control how your employees get to and from work, you can encourage them to reduce their commuting emissions by taking some of the following measures:

- **Encouraging walking and cycling commutes.** Walking and cycling to and from work is one way to eliminate GHG emissions from a commute while also getting some exercise in. (An active lifestyle is good for productivity and mental/physical health—great attributes in an employee!) Employers can encourage these commutes by providing information on local bike paths and walking routes. Safe bike storage, showers, and changing rooms will make employees more likely to want to walk or cycle to and from work. Partnering with local bike shops for a direct line to affordable (and perhaps quality secondhand) bikes is a smart move. And some companies further incentivize cycling via discounted rates on gear and/or interest-free loans that can be taken out of paychecks pretax.

- **Offer public transportation loans, discounts, and/or passes.** If your company is accessible via public transportation, offering discounts or loans on travel passes is a simple way to encourage employee use. Stipends can come out of an employee's pay pretax or be handed out as an additional employee benefit, via gift cards or discount codes for online purchases. Some transportation systems may offer group rates on passes, making it easier to pass these benefits on to staff. Also, many cities offer subsidy programs—don't leave money on the table!

- **Set up a company bus service or carsharing program.** Road travel is the largest emitter of CO_2 worldwide, so anything to minimize the number of vehicles on the road helps. If your facilities are not easily accessible by public transport, a shuttle service that runs employees between the nearest hub and your business could be worth considering. Also, if your employees travel between multiple worksites within a workday, setting up simple communications systems that allow them to coordinate car-shares is an easy win for the environment.

- **Choose sustainable transportation methods.** Planning ahead can often help you find sustainable means of business travel, whether that's choosing hybrid or electric car/taxi services or booking non-stop flights with environmentally friendly airlines.

- **Offer hybrid or remote work options.** Removing the need for a commute altogether, even if it's one day a week, both lowers your company's carbon footprint and can also save money spent on heating, lighting, water, and stocking your office with supplies. Allowing for asynchronous work and flexible hours can help your employees commute during off-peak hours, reducing traffic congestion and emissions from idling vehicles. Planning team meetings and on-site events so they line up on the same day or consecutive days is a great way to ensure that any energy use is optimized—for a full office, rather than a half-full one.

- **Encourage reuse.** For every item an employee reuses, there's a theoretical item out there that they didn't buy new. So, if an employee adds up all the things they reused instead of buying new, that's all carbon emission avoidance. Buying new or used is the single most impactful decision an employee can make for the world *and* your collective bottom line. Employees in all roles and at all levels of the organization should be encouraged—even incentivized—to find ways to reuse.

So far, many of these eco-friendly solutions have been smaller, and more tactical. Now, we're going to move to more strategic sustainability methods that have a larger, more sweeping impact.

Take Advantage of Carbon Offsets

Carbon offsets are a chance for individuals and companies to invest in carbon-reducing projects across the world to balance out their carbon emissions. A common criticism of carbon offsets is that they're an easy way for those with the financial means to clean their environmental conscience to do so, without changing their habits. We'd reframe them a last resort. After you've taken all the possible steps you can to reduce carbon emissions, carbon offsets can balance those you haven't been able to eliminate, yet.

The carbon-reducing projects that carbon offsets support usually try to reduce future emissions in developing countries. Others are more literal, planting trees to soak up actual CO_2 in the air. No matter which type of offsets you purchase, it's important to do your due diligence to make sure they're from a reputable source. Companies often purchase carbon offsets aiming to reduce total GHG emissions by an amount similar to their current GHG footprint; in this way, they can keep working toward net zero as an organization, knowing that any offsets they'll need to purchase in the future will be reduced. Individuals, on the other hand, tend to purchase carbon offsets to neutralize the impact of a specific activity, like a flight.

As of now, there isn't a viable solution for the entire aviation industry to reduce carbon emissions to zero. (Despite claims of some larger companies like Delta, most smaller airlines are unable to make this type of commitment.) That's why more than 50 airlines offer carbon offset programs for their passengers, either integrating offset purchasing into the online ticket buying process or sending customers to a third-party offset provider. But will this really move the needle fast enough, when weighed against the gargantuan carbon footprint of the industry as a whole?

One of the cases we're trying to make is for governments to recognize reuse as a (potentially preferrable) method for companies to accumulate carbon offsets. (Justin is specifically working with lobbyists and representatives on this effort.) After all, reuse is the most effective form of carbon offsetting—in fact, unlike other offsetting methods, there's a direct correlation between reuse and a decreased carbon footprint. Sticking with the aviation industry to illustrate our point, of the approximate 5,000 airlines across the world, only 300 are International Air Transport Association (IATA) members—but these IATA airlines account for 83% of air traffic. If all of them worked together to participate in a reuse program, it'd do a lot more good than Delta alone working toward net zero. Consider the process of plane manufacturing itself. Even though there were more than 6,300 aircraft in storage in 2023,[120] Boeing is forecasting the demand for nearly 44,000 new airplanes through 2043.[121] Think of all the virgin material extraction and energy usage required to manufacture these new planes (not to mention all the materials and energy already put into use to create the planes currently sitting in huge hangers and warehouses as wasted value). What if some of the materials and/or parts from those stored planes could be used in the new ones—or, if some of the stored planes could be refurbished and put into action again? And what could that do to support a growing technical service and support industry. It could make a massive difference.

In short, getting donations from passengers to offset carbon won't move the needle, but getting all planes into a reuse program from the day they start flying will make all of them more efficient, ensuring that we get as much value from them as possible.

Choose Your Partners Strategically

Choosing to work with other sustainable companies—whether as suppliers, vendors, partner organizations, customers, or in another capacity—can both help overcome boundaries and create relationships based on shared goals and a spirit of cooperation. Partnerships where all parties are focused on moving to circularity can share valuable learning while facing the challenges that come with change together.

Moreover, even within large organizations, different departments or brands can act as smaller companies connected by a portfolio owner. We've found that clients who operate in this siloed way can particularly benefit from one division undertaking CE methods and sharing them with other divisions. Indeed, when you work for a subsidiary of a much larger organization, a success story like one of the many we've covered thus far supplies a great reason to reach out to your parent or sister company, as many of these tools and techniques can be applied cross functionally.

More broadly, working with sustainable partners unlocks business value that compounds quickly. For one, promoting these partnerships gives you a competitive edge and a stronger reputation with conscientious consumers. And as sustainable organizations are also less vulnerable to the impact of environmental degradation, supply chain disruptions, and rising costs, partnering with them offers market resilience. Lastly, collaborative efforts toward sustainability fuel innovation, efficiency, and cost savings.

Cleaning product company Seventh Generation is one example of an organization focused on sustainability throughout the lifecycle of its products. According to their origin story on their website,[122] "Our name is inspired by an ancient Iroquois philosophy which instructs that 'in our every deliberation, we must consider the impact of our decisions on the next seven generations.' This credo guides us in every product we make, and every action we take. It inspires our belief in a seventh generation to come." They focus on full circle product design, using biodegradable ingredients and recycled and recyclable packaging, while holding their suppliers and manufacturers to the same standards— including a commitment to reducing their own carbon footprint.

Having built their business around these standards, Seventh Generation was acquired by Unilever in 2016 for an estimated $600–700 million.[123] An executive-in-residence at CASE, an award-winning research and education center based at Duke University's Fuqua School of Business, called this[124] "arguably the largest acquisition to date of a social enterprise… Just as important as the financials of the deal is what this deal may signal in the marketplace. The line between social enterprises and 'regular' enterprises may be disappearing. Social enterprises are now entering the mainstream."

We agree.

Get Started: Identifying Low-Hanging Fruit

Use this checklist—which includes ideas from this chapter as well as a handful of others we didn't get a chance to cover—to identify quick and easy ways you can get started with eco-friendly solutions in your workplace.

Office supplies
- ☐ Go paperless
- ☐ Copy and print on both sides of paper
- ☐ Use post-consumer recycled paper
- ☐ Use chlorine-free paper
- ☐ Switch to recycled ink
- ☐ Switch to recycled pens
- ☐ Eliminate single-use cups and plastics

- ☐ Encourage reusable and/or recyclable dishware
- ☐ Use eco-friendly cleaning products
- ☐ Order in bulk
- ☐ Use a coffee maker with a reusable filter
- ☐ Get rid of paper towels

Waste and resource management

- ☐ Get rid of individual waste bins, opting for communal waste stations
- ☐ Compost
- ☐ Donate your compost to a local community garden
- ☐ Use compostable garbage bags
- ☐ Swap out bathroom faucets for water-conservation faucets
- ☐ Add plants to your office space
- ☐ Set up monitoring systems on equipment
- ☐ Increase natural light
- ☐ Switch to energy-saving bulbs
- ☐ Turn off everything at night
- ☐ Unplug
- ☐ Use motion-activated light switches
- ☐ Use smart plugs and/or outlets
- ☐ Opt for renewable energy sources (solar and/or wind power)

- ☐ Optimize your heating and cooling systems
- ☐ Delete old emails and digital files to reduce the energy your organization needs from data centers

Reuse

- ☐ Purchase sustainable or reused furniture
- ☐ Purchase sustainable, energy-efficient and/or reused equipment
- ☐ Reuse, refurbish, sell or donate unused furniture and equipment

Carbon emissions

- ☐ Encourage walking and cycling commutes
- ☐ Offer public transportation loans, discounts, and/or passes
- ☐ Set up a bus or shuttle service
- ☐ Set up a carsharing program
- ☐ Choose sustainable transportation methods for business travel
- ☐ Offer hybrid or remote work options
- ☐ Encourage reuse for individuals across your organization
- ☐ Carbon offsets—as a last resort

Partnerships

- ☐ Build relationships with sustainable partners
- ☐ Ask all business partners (vendors, suppliers, etc.) to agree to a sustainable code of conduct for doing business together

Food and drink

- ☐ Provide vegetarian or vegan meals in your cafeteria
- ☐ Provide locally sourced food for your employees
- ☐ Provide organic food for your employees
- ☐ Set up water refill stations

Strategy

- ☐ Institute a triple bottom line approach
- ☐ Set up Go Green Challenges across your organization
- ☐ Start a Green Team
- ☐ Organize a volunteer day, donating your time to eco-friendly organizations
- ☐ Celebrate Earth Day with volunteerism and/or educational activities (consider a "lunch and learn" event)

As you start to implement these small changes in your business, you're going to run across challenges. You might feel overwhelmed. In chapter 6, we're going to show you how you can take advantage of artificial intelligence to help mitigate any issues and ease your transition to circularity.

Chapter 6

AI and CE Integration

Merriam-Webster defines artificial intelligence as "a branch of computer science dealing with the simulation of intelligent human behavior by computers."[125] We understand that many of you may view AI with trepidation. Perhaps you've seen images created by or enhanced with AI that show people with too many fingers on their hands, or heard about McDonald's botched attempt to use AI for taking drive-thru orders—including the customers who tried and failed to stop the AI from adding more Chicken McNuggets to their order as the total reached 260.[126] The AI Fails subreddit[127] has 15,000 members and counting. It's fun—and validating—to see robots mess up.

We understand this mindset. And we wouldn't use AI to, say, write this book. That's a job for a human mind. We've also heard the talk about AI being environmentally problematic, due to the impact of the data centers they're housed in.[128] However, since AI is here—and here to stay—when we're talking about developing effective circular strategies, AI can save a ton of time and, with straightforward tasks, reduce the potential for human error. And these are opportunities worth considering.

What Is the Potential Role of AI in the Circular Economy?

The Ellen MacArthur Foundation was ahead of the curve when, back in 2019, they published a paper titled *Artificial intelligence and the circular economy: AI as a tool to accelerate the transition.*[129] Google and McKinsey & Company collaborated on the project, which suggested that AI could help scale the transition to CE, specifically in three areas: product design, operations, and infrastructure optimization.

Product design: AI's potential application in designing circular products—i.e., using the best possible components and materials—lies mostly in assisting and accelerating the *human* design process. Training AI to help generate, test, and refine designs could improve outcomes, reducing the human tendency to lean on favored or known options. AI could also suggest materials that are as sustainable as possible, for example focusing on what's available locally and what will retain the most value as it moves through the lifecycle. As part of all these efforts, there are huge amounts of data on existing materials that can take a long time for humans to compare, contrast, and test; AI can assist with rapid analysis and feedback loops for testing.

Operations: Operating circular business models involves aligning all sorts of business functions around CE principles. Decision-making can be sped up and improved when informed by product and customer data analyzed by a powerful, well-trained AI.

Infrastructure optimization: Finally, ensuring circular loops of materials and products requires the best possible recovery level for reuse, repair, remanufacture, and recycling. Sorting and separating mixed streams of recovered materials and products is a great use of AI, and visual recognition techniques can train AI to sort waste. For example, ZenRobotics was able to train waste-sorting robots to an accuracy level of 98% across all sorts of materials streams.[130]

AI has incredible potential to help move a handful of particularly environmentally damaging industries toward CE. Specifically, in chapter 4 we talked about the food industry's desperate need for sustainability—i.e., how for every $1 spent on food, society pays $2 in economic, social, and environmental costs. One of AI's big roles here could be to help "design out" food waste in the distribution process, with automated sorting and processing techniques. Predictive models can help match food supply and demand more effectively and determine the best way to redesign supply chains with a focus on sourcing local ingredients. On farms, AI solutions could guide the most effective implementation of regenerative farming practices, including best methods for crop rotation, and simulate field trials that would otherwise require entire growing seasons to execute.

The innovative Norwegian collection and sorting equipment manufacturer TOMRA was founded in 1972 with an initial focus on reverse vending machines, for collecting used beverage containers. Their business model was built on circularity before it was even part of the lexicon! Now, they enable the circular economy via both advanced materials collection and waste sorting systems—and by making use of AI to process food in a resource-conscious way. TOMRA has sensor-based food sorting and grading technology that evaluates food on factors like level of ripeness. AI algorithms work to sort products based on potential uses and ensure that food products are cut into consistent pieces to reduce overall waste. What's more: the data that TOMRA gathers with its machines can lead to efficiency improvements over time, including optimizing line settings for their customers and informing TOMRA's ongoing R&D to innovate and refine their own technological solutions.

The consumer electronics sector is another industry that could get a huge circularity boost from AI. In fact, estimates show that AI could help guide the industry toward CE in a move that would generate up to $90 billion of top-line revenue growth per year by 2030.[131] Specifically, AI could enable designs that would extend the use period of electronics, help capture and recover a significant part of the total e-waste market, generate material use efficiencies and reduce waste during the production process, and speed up R&D via rapid testing opportunities.

One company that's using AI to empower circularity in the consumer electronics industry is Reconext—an industry leader in boosting reuse of electronic devices, managing huge flows of used electronics. (Much of their groundbreaking work began under the name Teleplan, before that company was acquired by Clover

Wireless and Reconext was announced as the combined identity of the new resulting company.) Reconext's website states that they process 30 million items annually, and that 40% of Fortune 500 Tech & Telecom companies—such big names as Amazon, Best Buy, Dell, HP, and Microsoft—rely on their services.

Many electronic devices, or their components, can be reused, and Reconext focuses its R&D efforts on salvaging more and more existing parts and components at every turn. AI's role at Reconext lies in assessing the cosmetic condition of used electronics, a function known in the industry as "grading." Machine learning helps quickly, accurately, and reliably differentiate between types of marks on a device (e.g., between a logo and a speaker, or a scratch or dent)—a process that is not only more consistent and objective than human evaluation, but that also reduces the assessment cost per unit, increasing profit margins.

One benefit of AI is the ability to quickly and seamlessly combine previously recorded and real-time data from everyone in the value chain it has access to: ideally producers, manufacturers, suppliers, and consumers. Having touched on smart product design and supply chain optimization, a few more ways AI can thus contribute to a circular future *across industries* include:

- Analyzing typical product lifespans in design phases, and suggesting material choices in line with appropriate reuse and/or recyclability goals.
- Predicting carbon costs and finding ways to minimize carbon footprints in manufacturing processes.
- Taking a holistic view of a supply chain and suggesting ways to minimize transportation costs and emissions, as well as improve overall efficiency.

Additionally, we've discussed real-time equipment monitoring multiple times. Combining the data you get from these tools with the analytical abilities of AI—including access to data about similar machines, potentially in real time—can be used to improve predictive maintenance efforts. Again, this minimizes downtime and reduces maintenance costs.

Next, we'd like to share how we've made productive use of AI in our company.

How We've Implemented AI at REUZEit

Like most people, Justin was worried when he first learned about AI being available to the masses. But his curiosity quickly trumped his concerns and, as he began playing with ChatGPT, he realized much of his worry was based on simple assumptions he'd gleaned from the news, or fears that had been stoked by dystopic movies. These were the days when AI was still using data it had acquired from the internet of a year prior, rather than working with real-time information. After several days of writing increasingly large and more complex prompts, and doing tests like translating Old World literature into modern surfer talk, Justin began to see how AI is just a tool. That is, like any tool, it can only be as useful as your desired use for it, and some people are better than others at using AI.

Ryan's role at REUZEit includes streamlining the processing department—the heart of our company. Processing truckloads of equipment that almost always contain a range of item makes, models, manufacturing years, and physical conditions is naturally time consuming—and requires the technical expertise to handle quality checks, manage professional photography, price each item for market, and so much more. We'd built a network of processors and professional writers to help us manage selling equipment online. The processors researched each piece of equipment, looking up the make/model, comparing the item on-hand to similar models, highlighting features and benefits, describing any known issues, and researching the most competitive price for the listing. It was a tedious process replicated across every piece of equipment—a clear bottleneck in our scalability. Furthermore, the technical writers worked part-time, from home, to write the thousands of different listings required to keep items moving out of our storage facility and back into productive use. In short, we make an expensive investment in every item we handle. We wanted to reduce this downtime, both to keep our assets liquid and to keep the circular loop in motion.

Ryan's initial refinement in our operations was to identify valuable, high-demand items that both require and merit additional time in processing. But even after he combined components into larger systems and developed a processing production line of sorts, it still took anywhere from 30 to 45 minutes to process most items. While we'd try to cut that time down to 10 or 15 minutes for cheaper goods, much of this work was manual: recording make and model, serial numbers, doing online competitive research, copy-pasting information, shopping for prices, and writing and posting a listing. Justin thought the best way to scale this work might be to outsource some of the work (as he'd done with a remote technical writing team he'd recruited); however, we soon realized that training and relying on freelance writers to handle this work—work that was central to the business' revenue—was neither cost effective nor scalable.

In late 2023, we hired Andrei, a real CTO, to take over these responsibilities. On day one, Justin started hounding Andrei about his idea to use AI for the processing department's heavy lifting. To his credit, and to prove to us that we probably should've hired an actual CTO sooner, Andrei quickly developed an incredible platform that took our business to the next level. Today, our use of AI at REUZEit is simple, but effective: we use it to handle the bulk of our reselling business functions. Essentially, AI helps us write online product listings and determine market value—all with human oversight.

It's amazing how fast AI is evolving. The first half-step we took toward automation was consulting with an algorithm developer to design an automated repricing formula that could help us price all our items based on several factors that we defined. But all of this is completely gone now. With the help of a particular set of AI tools—tools that weren't available and fully functional until 2024—we've transformed our seemingly unsolvable bottleneck into a competitive advantage overnight.

Now, our processors simply put an item on a table. An HD camera spins it around, recognizing the make, model, and year, then confirms that information with a database. Then, the AI goes even further. It reads our quality inspection and researches the internet to create an accurate, marketable listing, one that is ready to post to all our sales channels. It even considers our past transactions to estimate price, provides weights and dimensions, and gives us shipping classes and specific Harmonized Shipping Codes for each product. Essentially, all our processing team must do now is take photos (used for both the listing itself and the AI's analysis) and test the items, giving them the time needed to inspect and evaluate the quality of every piece of equipment we move—a human task.

On that note, we want to point out that AI is just a tool for writing and basic research. As it cannot discern what source(s) to trust, human processors need to check the data for AI bugs, and we're always tweak the AI prompts to improve its accuracy. However, in our opinion, AI has its time and place, and we are grateful for how it has helped us upgrade our workflow and efficiencies. The proof is in the numbers. Whereas before processing high value, complex items would take about 45 minutes, with the help of AI, it now takes about one minute.

Moving forward, we're anticipating a future state in which we can use AI to see the values of used items in real time and analyze the equipment vis-a-vis the market, pairing up assets with the person or lab who needs them. AI may also be able to one day track data that predicts not only an item's end-of-life, but also find its next supplier and/or buyer before a maintenance manager even notices an issue. All of this minimizes downtime and improves both efficiency and value extraction from the machines we—humans and the earth—have already paid to produce.

Digital Technologies Paving the Way for AI and CE Integration

A suite of digital tools that combine technology, data, and human creativity offer pathways toward integrating AI with CE—in pursuit of innovative, sustainable business operations. Below are a few of the most important. [Note that many of these technologies are used concurrently to achieve maximum efficiency benefits.]

High-performance computing (HPC) enables the quick and accurate processing of massive amounts of data. HPC systems are big groupings of computers, specialized software, and high-speed networks that collect tons of information. They provide the power needed to run AI and machine-learning (ML) models at incredible speeds. They also make it easier for organizations across industries to have access to analyses they never would have otherwise had, enabling them to solve complex problems and develop new products and services to meet identified gaps in the market. For example, Amazon Web Services (AWS) and Good Chemistry, a Vancouver-based computational chemistry company, used HPC to simulate and discover the exact compound that would destroy PFAS, the "forever plastic" molecule that has been found in water systems and the human body and is known to cause health problems.

Second, **big data analytics** offer the best path to building a comprehensive understanding of the impact of circularity. Data that establishes baselines, evaluates the effectiveness of interventions/new strategies, and measures outcomes will be critical to proving out CE's success. Some of the most common key performance indicators (KPIs) used to measure circularity—that is, to help companies see where they are on the path to circularity and identify pain points for improvement— are tracked via big data, including resource efficiency, recyclability, waste reduction, renewable energy use, and product design. Such data can be key for driving individual organizations' strategies. And, in aggregate, it can be used to argue the case for more companies to adopt CE strategies, and potentially for lawmakers and regulatory bodies to establish appropriate, actionable, and measurable industry standards and policies. (More on that in the next chapter.)

AI and ML algorithms are also fantastic tools for real-time monitoring of energy use and waste, frequently analyzing data for innovation, forecasting, and process optimization. One example of this is EDF Energy in the UK, a company that uses AI to monitor real-time use of their low-carbon electricity resources, enabling customers to downshift their usage for a discount on their bill (or even to make money) during times when the power grid is being stretched.

Next, the **Internet of Things (IoT)** represents the digital world's extension into the physical world—connecting tangible things to the digital world by software, sensors, and/or the internet. Benefits include the seamless, automatic exchange of data on customer use, item condition, and even material status points like moisture or temperature. (Note that this data is shared machine-to-machine, requiring zero human communication, intervention, or reporting.) This data can then inform CE strategies related to where an item is in its lifecycle based on comparative models, as well as predicting material flows that guide efficient resource management.

Lastly, **additive manufacturing** (another way of saying 3D printing) uses technology to minimize waste in the manufacturing process. It also allows for simple customization, on-demand production, and quick iteration without the need for developing custom manufacturing processes and tools like those needed for injection molding.

So, if the way has been paved for integrating the power of AI with CE efforts, where do we go from here?

Introducing AI to CE

CE is all about preserving product and material value for as long as possible, and AI can be applied at each step in shifting from the linear model (take-make-waste) to a more informed six-step process: think, take, make, use, take back, reuse.

One of the biggest differences between the linear and circular models is the introduction of a step before the traditional "taking" of resources—referred to by some as "think." And it's one of the biggest challenges of circularity. AI can draw on big databases, helping companies at the design stage to source the most sustainable materials and suggesting ways to combine them to create high-quality products that are easily dismantled and ideally reused, remanufactured, or upcycled. One example: food chemists at Chilean startup NotCo used AI to suggest ingredients for new vegan alternatives to dairy products with big carbon footprints. A human might not have thought to use pineapple and cabbage extracts in a vegan milk—but AI did, and it worked.[132] (Of course, humans had to be on hand to taste-test the resulting recipe!)

AI can also maximize the sustainability of resource extraction in the "take" process—by suggesting the best areas to mine or monitoring soil health for ideal crop rotation patterns and yields. Then, to "make" products as efficiently as possible, AI can monitor manufacturing processes for waste avoidance—in terms of machine or raw material usage, or even by reducing the quantity of packaging materials needed to keep a food product safe and fresh.

AI guides make a best-case scenario of "use," too. Smart washing machines are a great example. Once loaded, they can sense the size of the load and run a cycle that is the most energy efficient in terms of water and electrical usage. Some can even tap into the energy grid and suggest an optimal start time based on real-time electrical supply and demand.

Finally, we've discussed reverse supply chains many times, and establishing a strong "take back" process is one of the biggest hurdles most organizations have in the shift to CE. Once again, AI can facilitate this transition. Using optical diagnostic capabilities, AI can assess whether a recovered product is fit for resale, ought to be donated, could gain new life through repair or remanufacture, or if it's time to upcycle or recycle it. When it comes to "take back" and "reuse," to help close the loop as efficiently as possible, REUZEit is creating a visual dataset and connecting the metadata—applying AI to assist human diagnostics/reconditioning, as well as efficient processing and resale of equipment and other products.

Ready to consider how AI might improve your business workflows, efficiency, and sustainability measures?

Get Started: Map Out Your AI Journey

As we mentioned earlier, some people are better than others at using AI. Part of that is skill, much of it is practice, and most of it is being deliberate about your use of and for the tool. Here's our step-by-step guide on how to introduce AI into a business, so that you can be smart about things from the beginning.

The first thing to do? Learn about AI's capabilities and limitations. AI is not a magic bullet that can solve every business challenge on your list, and understanding the tool is the best way to begin determining how to use it. ITRex, an enterprise software development company that includes AI consulting at the top of their list of services, suggests[133] that AI works well for a specific handful of tasks across industries and use cases:

- Scheduling
- Forecasting
- Process enhancement and automation
- Resource management and allocation
- Reporting
- Cybersecurity management

[To be clear, its uses aren't explicitly limited to these categories, and AI continues to evolve, offering more opportunity to put it to use every day.]

Humans should always oversee AI, particularly during implementation and algorithm adjustment processes. ITRex notes that AI is limited in the following areas:

- Generating creative content (see: why we didn't use AI to write this book!)
- Coding complex software systems
- Making judgements and ethical decisions
- Coming up with innovative, creative solutions that don't already exist (because AI is limited to working with data on what *does* already exist)

The second step in introducing AI to your business is defining your goals. First, identify the problems you're trying to solve with the help of AI. This involves internal meetings to assess needs and gain buy-in. Then, make sure you establish clear objectives that allow you to measure the hard and soft ROI of your investments. Hard ROI can include things like time savings and labor/operational cost reductions (as with our use case at REUZEit). Soft ROI measures can include customer retention based on improved experience and employee satisfaction from automating monotonous tasks.

Step three involves evaluating your organization's AI readiness. You'll need talent to help you develop and implement AI, whether that's in-house IT specialists or third-party experts. Regardless, you'll need to train all employees who will work with any of the functions that AI touches so they can do their jobs effectively. Next, you need to understand if your strategy requires SaaS tools or custom software, which have their own budgetary implications, as do the computing and storage resources AI requires. Plus, AI needs data to operate, so you'll want to ensure you have access to all the data you need to get the best results from your efforts.

The fourth step is to begin integrating AI into select processes, starting small and learning as you plan for scale. Keeping bigger picture goals in mind will help you ensure that you're testing for the right things, and that you'll get a quicker payback on your investment. ITRex suggests a proof of concept should give you results within three months; if it doesn't, stop and try another scenario.

As you move toward larger-scale implementation of AI in your organization, you can use what you've learned to achieve AI excellence. Setting up systems around secure and efficient data management, with integration for algorithm training, will help you achieve ROI faster. Also, creating a central hub for employee training and information sharing will encourage other departments to try out AI—and could get the C-suite excited about further investments.

As fun as all this talk around playing with AI is, it's time to move on to a drier—but no less important—topic. In the penultimate chapter, we're going to discuss the rules and regulations around CE that are in place now, and those that would help move the needle to where it needs to be. Stick with us, because this stuff is critical!

Chapter 7

Mastering CE Standards and Regulations

We'll be blunt: international standards for the circular economy aren't up to snuff. Yet, some governments are taking steps to create policies and design regulations that will improve outcomes, and we'll look at those in this chapter—as well as the barriers that are holding many organizations (and nations!) back. We'll end with our recommendations for specific standards—grounded in our expertise in the space—to encourage businesses, worldwide, to make the changes our world needs.

Current State

Because there aren't many rules and regulations for those of us operating explicitly in the circular economy, it can sometimes feel like the Wild West. So, one time, REUZEit almost unintentionally broke the regulations of a federal US agency: the Drug Enforcement Administration (DEA).

During a site closure, our client asked us to remove several truckloads of equipment from pharmaceutical production lines. We're talking machines that are highly regulated, because if they were to end up in the wrong hands—like, say, those of a cartel—they could be used to manufacture illegal drugs. Usually, when we remove large equipment from a facility, we try to find a buyer in the region, to reduce the transport costs that come with moving such huge items. And at this point, as we didn't have a lot of space in our warehouse, we thought it would be a good idea to sell this pharmaceutical equipment in-place and ship each piece to the buyer, directly from the project site.

Luckily, we happened to talk to one of the scientists who ran the lab before taking any action. He was shutting down equipment in the final days of operation while we prepared rigging and construction. He told us that the DEA wouldn't allow us to resell this equipment to anyone. Fortunately for us, he was wrong, otherwise we'd be sitting on a ton of giant, seemingly valuable equipment with our hands tied. However, his warning did tip us off to the fact that we had to go through several procedures and checks before the DEA would grant us *permission* to transfer ownership, so we could ship the equipment. After a couple weeks' delay, everything was fine, but we dodged a bullet. If we hadn't gone through the proper steps—which we'd only realized were necessary in the eleventh hour—there could have been massive fines levied against us, our client, and even the shipping company we used.

Alongside the massive benefits of the circular economy, there is the risk that certain restricted equipment could be reused for nefarious purposes. To continue avoiding sending useful equipment and valuable materials to landfills, centralizing regulations around reuse would make it easier for companies operating in the space to keep things on the up-and-up—rather than having individual players trying to ascertain which regulations apply to which industries and which specific materials or products.

Our hope is that this chapter will both lend guidance for developing specific CE and SAM programming within your company—including ISOs that will get you started—and educate you on the current status of CE, giving you vocabulary to use as the conversation relates to broader policy concerns.

Discrepancies in the Circular Economy

"Policymakers with a holistic vision, such as zero-waste or a low-carbon economy, and the capability of identifying innovations in search of industrial applications, can exercise an important pull function through focused research programmes and procurement specifications."
— Walter R. Stahel, *The Circular Economy: A User's Guide*[134]

Some countries are excelling in codifying CE regulations while others are lagging. Many countries fall somewhere in between.

Across Europe, a few measures have been taken. In 2015, the European Commission approved a 54-measure action plan to "close the loop" of the product lifecycle, with a focus on five sectors. Then, in late 2019, the European Green Deal was presented and, as part of that, the Circular Economy Action Plan was approved in March 2020. The focus of this newer action plan was on design and production.

To kick things off, and as Europe is leading the world in circular economy practices, we're going to start by looking at a handful of European nations (addressed alphabetically and boiled down to summary for the sake of expediency) and move on to other parts of the world from there.

Belgium[135]

The term "circular economy" is well-known and used throughout Belgium. In fact, Brussels was established as a pilot city for the United Nations Environment Programme (UNEP) to track the transition to a circular economy, including measures of jobs created from the shift. They started ahead of the pack, too—in 2016, almost 77% of waste in Belgium was recycled, ranking it second-best within the EU (behind Italy).

As a nation, Belgium has committed to UNEP's 2030 Agenda, with the Flanders region setting a goal of having a fully circular economy by 2050. Specific CE-related initiatives across the country include multiple "green deals" (voluntary agreements between private, public, and government partners) such as Flanders' Green Deal on Buildings and Construction, launched in

2019; the "Be Circular, Be Brussels" initiative that engaged companies to submit their circular projects; Ghent en Garde, a sustainable urban food policy launched in Ghent in 2013; and other food partnerships and action plans in Flanders and Wallonia. The building and construction sector in Belgium is making some of the most impressive strides toward circularity.

France

France's Energy Transition Law (2015)[136] and their roadmap for circular economy[137] (2018, identifying 50 measures to get to a 100% circular economy) both demonstrate the country's favorable policy toward and clear speech on circularity. In February 2020, France also approved a law against squandering resources,[138] including measures on reducing waste, improving consumer access to information, and promoting reuse.[139]

Germany

The Organization for Economic Cooperation and Development's (OECD) "Effective Carbon Rates 2021" report[140] highlights Germany's leadership in implementing effective carbon pricing strategies and environmental regulations, which are pivotal for supporting CE frameworks, particularly through reduced carbon emissions and material use. Germany also often leads Europe in producer responsibility laws for the electronics and automotive sectors, regulations that align with CE principles to reduce waste and improve resource efficiency.[141]

Some of the sustainable policies that have been adopted in Germany include the Sustainable Development Strategy, the National Programme for Sustainable Consumption, and the Resource Efficiency Programme. The WEF Circularity Gap Report from 2021[142] commends the Resource Efficiency Programme in particular for its contribution to advancing circularity, via its focus on emphasizing material reduction and recycling targets. However, the report also critiques limitations the country faces integrating CE measures across all sectors of production and consumption. Overall, Germany is a clear leader in waste management and resource efficiency but has yet to address circularity in a holistic way.

Italy

The Italian Foundation for Sustainable Development created a Circular Economy Network 2020 Report that gave Italy the top position in Europe for circular efforts;[143] however, as with any self-granted award, it's fair to be skeptical. Their 2020 budget law[144] included some measures to comply with the EU Green Deal,[145] including establishing an investment fund earmarked for sustainable initiatives. So, some policies and regulations are certainly in place. Yet the need for these types of investment funds demonstrates there is room for structural changes that will aid in the country's shift to circularity and resource efficiency.

Luxembourg

Luxembourg presented their "Third Industrial Revolution (TIR)" strategy in 2016.[146] The circular economy featured prominently within its objectives, and one of their stated visions for 2050 was: "Luxembourg will be the first circular nation, where new business models based on the product as a service principle become standard." The specific strategic measures they planned to use to get there were as follows:

Policy and regulation:

1. Implement a taxation system that places the nation as the EU circular economy leader
2. Government plays an active public role in promoting circular economy

Educational:
3. New educational curricula for circular product design and production
4. Large scale awareness campaigns for the public

Business model innovation:
5. Promote circular design of products, new business models and circular supplies
6. Engage the farming community in producing wind and solar PV power as "cash crops"
7. Bring carbon back into the ground while promoting local organic products

Furthermore, Luxembourg hoped to center their efforts on six main sectors: food, manufacturing, construction, energy, transport, and finance. They also work with neighboring countries to further everyone's path to their separate and collaborative CE goals.

The Netherlands

The reason one of our offices is in the Netherlands is because it's one of the leading countries, worldwide, for the circular economy. Like Luxembourg, the Dutch government has a goal of being 100% circular by 2050. (Let's see who can get there first!) Their report "A Circular Economy in the Netherlands by 2050,"[147] published in September 2016, set out a roadmap to get there—including a government-wide program to streamline other programs and policy paths the country had been pursuing up to that point, including the Biomass Vision for 2030, the Raw Materials Memorandum, and "From Waste to Resources and Biobased Economy."[148] The government has demonstrated action toward the plan's goals, including approving transition agendas for five sectors (in 2018) and laying out specific projects (in 2019). Most CE efforts in the Netherlands thus far are effective at the local level, especially in sustainable energy; it'll be interesting to track how nationwide adoption via these policies impacts larger-scale change.

Portugal

Portugal adopted the "Action Plan for a Circular Economy in Portugal 2017–2020,"[149] which led to an increased interest in and awareness of resource management. The plan set out goals for 2050 that included carbon neutrality, a focus on research and innovation, and inclusivity, with efforts toward collaboration and safeguards against risk and volatility.

Spain

Spain has held policies that promote sustainable development and construction, ecological design, and recycling for a relatively long time. But in June 2020, they introduced Estrategia Española de Economía Circular: España Circular 2030[150] along with a series of three-year action plans for implementation. Goals include a 30% reduction of material consumption, 15% reduction in waste generation, and carbon emissions below 10 million tonnes of CO2. Unfortunately, before this plan was introduced, only 30% of Spanish companies claimed to have initiatives related to improving their sustainability, and they were only allocating 12% of their investment resources toward such projects.[151]

The United Kingdom
In February 2024, the United Kingdom announced they planned to establish the world's first UN-backed

International Centre of Excellence on Sustainable Resource Management in the Circular Economy (ICE-SRM). The center opened in April 2024 with a vision to "develop sustainable approaches to the circular economy and resource efficiency to enable carbon reduction and the transition to a greener future."[152] The center built on the successful four-year, £30-million National Interdisciplinary Circular Economy Research (NICER) Programme, an investment established in 2019 by the Department for Environment, Food and Rural Affairs' chief scientist, under the UK Government's Strategic Priorities Fund. Five institutions plan to provide research and industry expertise that will inform effective innovation and policy recommendations, with a focus on measurable long-term impact.

It's good that the UK is taking an active leadership role in CE because, unfortunately, its economy has a long way to go to reach their pledge of net-zero GHG emissions by 2050. The Circularity Gap report for the UK, published in 2024 by Circle Economy in collaboration with Deloitte,[153] found the UK population consumes 15.3 tonnes of materials per person annually, which is 20% more than the world average. Only 7.5% of these materials are used again, and, what's more, 90% of the

UK's material use comes from virgin sources (80% of which are extracted abroad). As a nation with tons of old buildings and a cooler climate, energy use is also quite high. Fortunately, for an economy of such a scale, small changes could make a huge difference. Circle Economy estimates that adopting more circular solutions in construction and manufacturing, as well as shifts in lifestyle choices, could cut UK material use by 40% and decrease its carbon footprint by 43%.

The United States

According to the University of Michigan Center for Sustainable Systems, the US remains one of the largest consumers of natural resources, across the continents,[154] with significant room for improvement in terms of sustainability practice. Underscoring the nation's outsized resource demand, US per capita material use reached 23.5 tons in 2022, 33% higher than the OECD average.[155] (To put that into perspective, a 40-foot shipping container packed with goods weighs anywhere from 22–30 tons, so each person in the US has an annual consumption that can be roughly equated to a fully packed shipping container.) Raw material use surpassed 3 gigatons in 2020, with construction materials such as stone and gravel accounting for nearly 75% of the total.

Encouragingly, recycling and composting diverted over 93 million tons of municipal solid waste from landfills and incinerators in 2018, demonstrating progress in waste management. (This effort also supported over 681,000 jobs and generated $5.4 billion in tax revenue in 2012, showcasing the economic benefits of a robust recycling infrastructure.) Alarmingly, however, only 8% of plastics disposed of in 2017 were recycled, while 2% leaked into the environment as microplastics, contributing to pollution and potential health risks. And the declining use of renewable materials (down from 41% to 5% of total material weight over the last century) highlights a shift toward *more* energy-intensive industrial production. Addressing these issues requires scaling successful models, such as aluminum recycling (beverage cans now use 73% recycled material), significantly reducing GHG emissions compared to virgin production.[156]

A 2018 report from Circular CoLab, *The State of the Circular Economy in America,*[157] analyzed 202 US-based CE initiatives to assess trends and establish a baseline for actions already in play. It found that the most popular theory of change being used in the US was "waste as a resource," with almost 40% of initiatives including this concept within their business model. Other key findings included the following:

- CE solutions exist in America, but many aren't referred to as such. For example, Goodwill has been operating in reuse since 1902 but does not explicitly cite CE in their mission.

- Less than 8% of companies in the study that sold physical products offered take-back options, and less than 40% of consumer goods companies offered easily accessible information on proper disposal options for their products. (The study authors recommended more focus on closing the loop.)

- Lastly, few organizations were considering radical alternatives that completely broke free of the traditional linear system—a necessary hurdle in transitioning fully to a circular model.

By aligning material consumption with CE principles, promoting renewable materials, and enhancing recycling systems, the US can achieve both economic growth and environmental sustainability. This dual focus is critical as resource pressures and environmental concerns escalate.

Africa

Out of 54 African countries, 50 have some degree of waste policies, strategies, or legislation in place, though implementation and enforcement vary widely. In a 2020 article published on CircularEconomy.earth[158], Kweku Attafuah-Wadee (a waste management consultant and Director at Resource Transformation Ghana (RTG) Ltd.) and Johanna Tilkanen (a project manager at the Environment and Society Programme of Chatham House) stated, "Repair, refurbishment, re-manufacturing and utilizing waste as a resource, particularly in agriculture, have traditionally had a significant role in many African economies." South Africa, Kenya, Namibia, and Tanzania have comprehensive plans that center CE. However, concepts like CE and the green economy (GE)—a UNEP framework that could feed into CE—and how to execute on them are vaguely recognized across the African continent.

In their 2019 chapter "Accelerating the Transition to a Circular Economy in Africa—case studies from Kenya and South Africa," authors Peter Desmond and Milcah Asamba elaborate:[159]

> "Different countries in Africa are at different stages of implementing GE with some countries only integrating aspects of it and others, like Ethiopia, Kenya and Rwanda, have put in place an overarching GE strategy. The legal and regulatory framework to foster GE is still in its infancy stage in most African countries and mechanisms to realise the transition to GE are not yet in place. The most promising markets tend to be those related to agriculture such as bio-trade, sustainable tourism and renewable energies (Klein, et al., 2013)."

African nations contribute to environmental degradation often because multinational businesses with worldwide value chains, concentrated in the North, rely on them for resource extraction. For example, China relies on rare minerals from DRC to produce smartphones, and Nigeria's oil and gas industry is one of the top 10 GHG emitters in the world. Also, increased municipal waste generation paired with inadequate municipal waste collection—with specific problematic waste streams like e-waste that average a mere 0.9% recycle rate across the continent—are major concerns for Africa. In many ways, CE's key principles—like reducing consumption, minimizing harm in resource extraction, and resource reuse/recovery—would benefit Africa's local economy and environment.

All this being said, one of the benefits of a less formal and developed manufacturing infrastructure is that an emerging economy (such as many of those in Africa) may be able to avoid resource-intensive linear economy practices in favor of more sustainable approaches from the get-go—i.e., leapfrogging over harmful models before they're ever put in place, and before the economy can grow to rely on them. This allows a country and economy to avoid expensive and time-consuming strategy shifts that require group buy-in and reallocation of assets, instead *starting* with a circular concept.

More future-focused work is being conducted by the African Circular Economy Alliance (ACEA), a government-led coalition of African countries conceived during the 2016 World Economic Forum "with a mission to spur Africa's transformation to a circular economy that delivers economic growth, jobs, and positive environmental outcomes." The ACEA is hosted by the African Development Bank (AfDB) and supported by the Africa Circular Economy Facility (ACEF), the latter of which became operational in 2022 and has a stated objective of "mainstreaming the circular economy as an inclusive green growth strategy to help African nations fulfil their development priorities while meeting the goals of the Paris Agreement, Sustainable Development Goals (SDGs) and the African Union's Agenda 2063."[160] The founding member countries include Nigeria, Rwanda, and South Africa, among 11 others.

India[161]

One of the biggest challenges the vast nation of India must overcome on their path to circularity is waste production and management. Their recycling rate is below 10%, and urban areas generate 42 million tons of solid waste annually. Other top hurdles include little awareness around CE, a lack of adequate infrastructure to support circularity, and regulatory barriers.

Nevertheless, unlocking CE's potential in India could have a staggering impact. The National Circular Economy Framework (NCEF) put out by the Confederation of Indian Industry (CII) and other leaders projected CE could have a value of $2 trillion in India by 2050, with an ambitious initiative to create 10 million jobs. A comprehensive approach that tackles regulatory reform, infrastructure development, and wide-scale awareness of the principles and benefits of circularity will be critical in realizing this potential.

The early good news? Some private sector initiatives, such as the youth-led movement Tide Turners and the Global Plastic Action Partnership (GPAP) demonstrate collaborative approaches to sustainability, as do India's efforts to enforce EPR guidelines and set goals for plastic waste. One business of note is Doodlage,[162] an Indian company that upcycles discarded factory waste and recycles post-consumer waste, engaging skilled artisans to create unique, high-quality clothing from cotton, silk, and wool.

Asia

Across the rest of the Asian continent, the status of the shift to circularity runs the gamut. As one of the largest economies in the world, many look to China as representative of the continent, and in fact, as far as circularity goes, China is a leader. The Circular Economy Promotion Law of the People's Republic of China went into force in 2009,[163] and the government has since implemented a wide array of policies and initiatives that encourage resource efficiency and sustainable production and consumption. Financial incentives include VAT exemptions and refunds.

South Korea is another leader, announcing a strategy in June 2023 aimed at reducing carbon emissions and stabilizing domestic supply chains across nine major industries, including those that involve tremendous energy usage and natural resource extraction.[164] They enacted important recycling laws and regulations back in the early 90s, including the Waste Control Act, which was revised in 1991 and introduced a waste deposit system for manufacturers, and the 1992 Act on the Promotion of Saving and Recycling of Resources, which, among other things, regulated single-use products.[165] In 2022, the Asian Development Bank Institute produced a report titled "Prospects for Transitioning from a Linear to Circular Economy in Developing Asia" which states that developing Asia accounts for as much as 53% of GHG emissions, and that 17 of the world's 50 largest landfills are in Asia.[166] And as the region continues to grow, waste generation becomes an even bigger issue to tackle: the World Bank estimates that Asia will generate 1.5 billion tons of municipal solid waste in 2030, increasing to 1.9 billion tons by 2050.[167] As we mentioned in chapter 4, cities are some of the biggest culprits in food waste, and 20 of the world's 33 megacities are in Asia and the Pacific, with the Asian Development Bank predicting this number will reach 27 by 2030. But perhaps we can look at the glass as being half full. Today, these cities may represent overconsumption and excess waste; yet, with a strong shift to circularity, they could be driving centers of sustainability.

In 2021, the Association of Southeast Asian Nations (ASEAN)—representing a region that is home to many quickly developing nations—adopted a framework for circularity[168] with five strategic priorities:

1. Harmonization of standards and mutual recognition of circular products and services

2. Trade openness in and facilitation of circular goods and services
3. Enhancing the role of innovation, digitalization, and emerging technologies
4. Competitive sustainable finance and innovative ESG investments
5. Efficient use of energy and other resources

Clothing is one of the most toxic manufacturing industries worldwide, and many of our clothing items carry labels showing they were made in China, Vietnam, Cambodia, Indonesia, Bangladesh, and other Asian countries. (According to McKinsey's State of Fashion 2024 report, 70% of the EU's textiles are manufactured in Asia. And according to the 2022 report, these brands had an estimated annual reach of $17 billion with an estimated growth of over 50% over the following five years.[169]) But exciting developments in upcycled Asian fashion brands are poised to bring about change. The OEC, an online data and visualization platform, estimates that 7 million tons of used clothes have been imported *into* Asia, most of which were upcycled into quilts, blankets, and other clothing. (2023 data[170] put the total value of world trade in used clothing at $5.13B, with the United States exporting $1.09B and Pakistan importing $335M worth.) Miuk Style, a clothing brand based in Central Vietnam, uses excess fabric to create traditional *ao dai* as well as other styles of traditional Asian clothing, including hanbok, kimono, and kurta. Designer and owner Xuan Thao has created outfits for renowned local singers and costumes for films.

Regulatory Barriers

Even in the EU, where perhaps the most progress has been made toward circularity relative to the rest of the world, there are still a handful of regulatory barriers that are slowing down the transition—ultimately holding back economic opportunities.

Some of the main regulatory barriers found by a third-party 2016 study[171] include the following:

- **A lack of clear, agreed-upon definitions and gaps in legislation.** For example, there aren't specific regulatory requirements for remanufacturing power tools, in part due to uncertainties in defining waste. The Waste Framework Directive (WFD) in particular contains problematic product definitions that result in only certified waste collectors being able to deal with certain secondary materials.

- **Unclear definitions of targets in legislation,** again in the context of the WFD.

- **The way that hard numerical limits are defined in current regulation.** First, there are inconsistencies. Second, much of the existing waste legislation is focused on quantity as opposed to quality of recycled materials. Registration, Evaluation, Authorization and Restriction of Chemicals (REACH) and Classification, Labelling and Packaging (CLP) regulations have been frequently cited as making it confusing to develop smart, effective reuse and feedback loops. None of this benefits reuse and secondary markets.

- **Lagging or partial implementation or enforcement of policies,** with the WFD proving problematic once again, along with the Exports and Shipment regulation.

- **Nations implementing legislation in different or even conflicting ways,** particularly in the context of directives and national action plans (when looked at in concert with the WFD, Basel Convention, and WEEE Directive).

- **Policies that conflict with each other because they were designed around discordant values,** e.g., with hygiene rules and food waste. More data could inform holistic, multi-faceted solutions that keep important hygiene standards in place without needing to sacrifice the potential value in food waste resources, for example.

In a broader sense, across the world, we can look at ESG standards and reporting and how growing interest in ESG companies as investment vehicles doesn't align with policy focus, or clarity. In 2024, T. Row Price conducted a Defined Contribution (DC) Consultant Study that reported "ESG integration is identified by most respondents (71%) as the best method of implementation within DC plans; however, the consultant and advisor community continues to view the evolving regulatory and legislative developments related to ESG as challenging."

Things aren't perfect, but the circular economy is still in its infancy. Think of how the Industrial Revolution spurred the need for regulation around things like workplace safety, child labor laws, and work hour limitations. As those policies demonstrate, industry was growing at an unhealthy pace, both for human beings and, as we now know, for the environment. Regulation kept things in check, to a point. Today, we international citizens and business leaders know more about scaling solutions successfully than we did in the 18th and 19th centuries—and doing so while moving the economy toward a more sustainable future is well worth the effort. Great progress toward useful and relevant standards, policies, and improved legislation is being made. Read on for information about a couple of shining examples.

The International Electrotechnical Commission and TC 111

In June 2022, the International Electrotechnical Commission (IEC) introduced TC 111[172] as a strategic business plan titled "Environmental standardization for electrical and electronic products and systems." The plan aims to accomplish the following:

- Prepare guidelines, basic and horizontal standards, and technical reports.
- Liaise with product committees about environmental requirements of product standards.

- Standardize activities worldwide.

There are 27 participating countries and 12 observing countries, including many EU nations as well as large economies (Brazil, Canada, China, India, Korea, Russia, Japan, the UK, and the US) and smaller countries (Colombia, Egypt, Malaysia, and Nigeria).

TC 111 has worked on such projects as determining specific substances in electrotechnical products (thus enabling safe and efficient reuse), creating a glossary of standardized terms, defining end-of-life information guidelines for recyclability, quantifying carbon footprint and GHG emission reduction, and more. It's incredible to see international organizations working together toward common goals. To have industry-wide standards, based on deep knowledge of the materials used and use cases across the world, is a fantastic baseline for any industry to use when building circular systems.

The ISO 14001 and 55000s and Asset Management Standards

Predating the new 2024 ISOs for CE by a decade, ISO 14001, ISO 55000, ISO 55001, and ISO 55002 are four documents that worked to establish a universal framework for asset management. (OK. We know, we know. It's a lot of acronyms and numbers. Talking about TCs and ISOs can start to feel like robot speak. But we can't overstate how pivotal the burgeoning field of asset management standards could be in pushing industries toward circularity. It's comparable to agreeing upon the internationally recognized rules for a sport *before* you start a game, or naming the clef, key, and meter for all the instruments in a symphony at the beginning of a piece of music. If we aren't all on the same page, then how can we play together, communicate with one another, and collaborate on something beautiful, or meaningful, or useful?)

Organizations can get certified around any (or all!) of these ISOs to demonstrate a commitment to sustainability, boosting market competitiveness in the process. We'll explain each, and why they matter, below. Think of them as sample rulebooks, if it helps you get through it!

ISO 14001
ISO 14001 is an internationally recognized standard for effective environmental management systems (EMS).

It provides a structured framework organizations can use to manage their environmental impacts; specifically, the committed environmental policies attached to ISO 14001 include improving resource efficiency, reducing waste, and complying with environmental regulations. (As you can probably gather, these sync right up with CE principles.)

One explicit mandate of ISO 14001 is extending environmental responsibility across the lifecycle of a product, considering everything from design to disposal—essentially, a circular approach. Another is continual improvement of EMS, which lines up with CE's principle to innovate and adapt at every stage of the process, always reaching toward nature's inherent regenerative cycles as a goal.

ISO 55000s (55000, 55001, and 55002)

The ISO 55000 series of regulations provide terminology, requirements, and guidance on implementing, maintaining, and improving an effective asset management system. As the Institute of Asset Management states, "We believe that asset management will increasingly become the best means of harnessing and integrating all the activities within an organization to achieve its purpose most effectively and economically." Of course, we couldn't agree more!

There are three standards in the series, and they're interconnected. ISO 55000 is an overview that introduces principles and terminology, which is necessary to developing a long-term plan that works for your organization's mission and values. ISO 55001 specifies the requirements for establishing and maintaining an asset management system, as well as improving it over time. Finally, ISO 55002 offers guidelines on to how to apply the requirements of ISO 55001 within your organization.

As we mentioned, additional standards falling under the ISO 55000 framework were released in 2024, with names that make their functions self-evident:

- ISO 55010 – Asset Management – Guidance on the Alignment of Financial and Non-Financial Functions in Asset Management
- ISO 55011 – Asset Management – Guidance for Development of Public Policy to Enable Asset Management

- ISO 55012 – Asset Management – Guidelines for Enhancing People Involvement and Competence
- ISO 55013 – Asset Management – Guidance on Data for Asset Management

The roadmap for implementing ISO 55000 is fairly straightforward. Begin by ensuring your leadership team is aligned on the value of certification, then bring any stakeholders and relevant teams into the fold. At that point, start assessing your organization's readiness; look at your current state and compare it against the international standards to see what you'll need to do to implement them. Next, you'll want to develop an asset management system by creating a policy, objectives, strategy, and plan. After that, performing a detailed gap assessment will help you move on to the following step, which is to bridge that gap! Finally, you implement the system, evaluate it, and finetune it. And importantly, don't be afraid to find that things don't work out entirely as planned. Look at this as a chance to continue learning, and to keep getting better.

Further Reading on CE and CAM Policies

We could talk about this stuff for ages, and there's probably space for another book to be written on the topic of this chapter alone, but we'll leave you here with a handful of publications and papers that you can seek out if you're interested in learning more about different CE and CAM policies, standards, and regulations.

- International trade and circular economy—Policy Alignment (OECD Trade and Environment Working Paper)
- United Nations Economist Network's "New Economics for Sustainable Development: Circular Economy"
- World Resources Forum's "Circular Economy—Policies and Learnings from Europe"
- European Commission's "A new Circular Economy Action Plan for a cleaner and more competitive Europe"
- Circular Asia's Government Policy and Regulation fact sheets
- Readout of the White House Circular Economy Innovation Roundtable

Get Started: Get ISO Certified and/or Establish Baseline Metrics

If possible, getting an ISO certification for your organization is a great way to throw your hat into the proverbial international ring of CE players. If that feels out of reach right now, perhaps laying out the specific barriers you're anticipating would be a good place to start. Craft a document. Look at policy, regulations, and standards within your industry, nation, and region. Identify what's holding you back and where you might find the push (or incentive!) you need to make a big shift.

Another thing you can do is pinpoint a suite of metrics that will help you establish a baseline of circularity. Despite the ISOs we mentioned, there are no set standards on metrics that measure circularity. So, at this point, it makes the most sense to look at what's out there and decide which measures work best for your organization. Then, start measuring and monitor progress. Set a plan to check back in on your numbers regularly, and report out. Establish an informal "green committee" in your organization to bounce ideas around. Once you start seeing your numbers improve (and you will), you'll likely have a case to make bigger changes in your department or company—because circular principles *help the bottom line.*

As another resource, consider this. In October 2020, Circle Economy and the Platform for Accelerating the Circular Economy (PACE) published a paper called "Circular Metrics for Business: Finding opportunities in the circular economy"[173] that offers ideas for how to get started with CE metrics. They bucket them into three categories:

- **Headline indicators** assess how circular your business is to help you understand how you can (and why you ought to) pivot your business toward circularity. An example metric is the overall circularity of a value chain, which could be expressed in percentages or unit of resources consumed per unit of revenue generated. The key here is to limit your metrics and to use them to target a desired (circular) end state, as opposed to the means you'll need to get there.
- **Performance indicators** help you understand what needs to be changed in your value chain to reach circularity. They focus on the production steps and material flows in your processes to provide insights into where you need to make changes. Example performance indicators include the recycling rate of a product, waste

generated in each step of the value chain (tied to a consistent unit or units), and the energy used in (and/or produced by) each step of the process.

- **Process indicators** help you determine how to bring about the changes you need to make to get to circularity. (I.e., the means to the end goal of functioning wholly within the CE.) These can be linked to everything from operational activities to human behavior. Examples include everything from a simple measure of the share of sustainable products in your portfolio to a more complex understanding of your customer's opinion of your sustainability efforts relative to those of your competitors.

As we move on from our look at policies regulating circularity—and the associated barriers—the final chapter looks at how to convert an organization to circularity in a way that will make everyone on your team embrace the effort. We'll also lay out the steps business owners can take to encourage others to run more sustainable businesses and lead more environmentally friendly lives—all without standing up on a soapbox. It's time to start making a difference!

Chapter 8

Next-Level CE: Becoming Part of Something Bigger

It's Justin, kicking off our final chapter by returning to a previous chapter in my life—one I've mentioned before.

After my first career as a financial analyst, I attended UC Davis's Master Brewer program. (For those who don't know much about the world of beer, aside from Doemans in Munich, Germany, UC Davis is the most renowned brewing school in the world.) I studied fermentation science and was certified by the Institute of Brewing and Distilling as a professional brewer. Today, about the only thing that remains of that time in my life is the D.Brew I can add after my name. Suffice it to say I have forgotten more about brewing than most people will ever know. But one memory that holds fast is the keg tap we had in the lecture hall, which had the added benefit of making my advanced degree much more fun to attain than most.

Because of these years of (drinking and) studying, I thought I knew about all the ingredients out there that could be used to make beer. But even I, Justin Andrews, D.Brew—and CEO of a company focused on reuse—never thought of turning leftover bread into a delicious fermented beverage. Luckily, Toast Brewing did.

Toast Brewing is an award-winning UK brewery. Established in 2016, they found a way to replace 25% of the main ingredient in beer, malted barley, with leftover bread. In doing so, they've also saved so many slices of bread that stacked up together, they'd be nearly five times the height of Mount Everest—an extra helpful reuse case, as about 44% of bread that we make as humans is wasted. And rather than keeping this proprietary ingredient and concept to themselves, Toast launched Companion, a brewing ingredient made from surplus bread that they're selling to other breweries. It's cheaper than virgin malted barley and has a lower carbon footprint. According to ReLondon,[174] "a craft brewery using Companion to replace just 5% of their virgin malt can save 1.2 tonnes of CO2 per year, while also preventing food waste."

Toast *started* with a circular model. And now, they're bringing more members of the beer industry into the fold. This goes to show that no matter how big or small your company or industry is, how young or old, circular practices have a place. And the time to implement them is now.

Company-Wide CE Enthusiasm: Start at the Top

As we've alluded to, it's key to engage all stakeholders throughout the process of transitioning to the circular economy because CE requires a holistic approach to be most effective. Getting a smart, dynamic team involved in the early stages of a strategic and operational transition will set you up for optimal success in the following ways:

- Forging relationships and fostering collaboration.
- Giving you the benefit of diverse perspectives in the earliest stages of planning. (You'll be able to make a roadmap based on collective knowledge and valuable insights that best serves all members of your organization and improves processes every step of the way.)

- Building a sense of ownership over the new system(s) for everyone involved. (Ultimately, because the circular economy is highly interdependent, trust must be built through collaboration, regular communication, and transparency. So why not start at the beginning?)

To foster enthusiasm for CE, highlighting business factors is a great place to start. And it's an easy case to make. First, as we've explored in depth, strategies to reduce waste and increase efficiency demonstrate organic productivity that helps the bottom line. Second, it's straightforward to convey ROI with the cost avoidance that goes hand-in-hand with reuse. Finally, capital recovery through refurbishment and resale can actually bring money *back into the company* as opposed to *paying* for services that will recycle equipment.

Once the wheels of new circular models have begun turning, it's pivotal that organizations measure and share the impacts of their efforts—both for learning/iterating and to celebrate successes.

Company-Wide CE Enthusiasm: Getting the Team on Board

OK—say you're spearheading the strategic CE talks within your company's leadership ranks. Let's get a little more tactical. Because encouraging employee engagement in CE initiatives begins with explaining the benefits of the circular economy—for your business and stakeholders as well as for future generations—you might host a town hall or make resources available on your Intranet or employee blog. (You could offer materials that explain how CE can help your organization reach sustainability goals, enhance your industry reputation, improve competitiveness amongst your peers, and provide new opportunities for innovation and collaboration. Case studies of success and data can illustrate the positive impacts of circularity—as well as show how dire the need for change is, which may resonate strongly for employees with children.)

Next, enable your team to apply circular practices in their daily work by giving them appropriate tools, training, support, and access to whoever on staff is best equipped to help them assess and execute their ideas. Then, involve employees by asking for their input and feedback. Why? Leadership can never see all the on-the-ground changes that are possible within their organizations. Encourage everyone to share ideas or suggestions, perhaps in a Shared Google Doc/Sheet or via specific, cross-functional teams or taskforces.

Additionally, monitoring and measuring initiatives using clear and relevant targets and reporting systems is necessary to evaluate your performance. This data will help you identify gaps and areas for improvement as well as inform the steps that will get you closer and closer to a closed loop model. And as all good leaders know, a little recognition goes a long way. Celebrating employee successes, acknowledging efforts, and offering rewards or incentives for particularly high-performing ideas and executions is always a great practice that will feed further innovation, and CE is no exception. There's no end to the ways you can bring this into action in your organization, whether it's through bonuses, career development, or special recognition programs or events. Success feeds off success, and using positive stories of circularity internally will likely inspire even more effort.

Lastly, joining networks or establishing partnerships with other organizations working toward circularity will give you access to best practices and the ability to learn from both the mistakes and accomplishments of others—thus accelerating the impact of your own efforts. These relationships could come in the form of peers, customers, suppliers, nonprofits, or academic partners.

The REUZEit Revolution

Speaking of idea sharing and social activation, at REUZEit we've embarked on a grassroots initiative to get more business involved in the circular economy. We call it the REUZEit Revolution.

How many industries can benefit from organic productivity, supply chain stability, and investment recovery?

Answer: all of them.

And how many people benefit from less pollution and accessible technology?

Answer: all of us.

The REUZEit Revolution idea is to empower *anyone* to pioneer the shift to circularity. To be an advocate for a commonsense approach to consumption, through reusing, repairing, and reallocating resources. To encourage others to extend the lifecycle of their products and equipment through reuse.

We hope that we can be a network for identifying businesses engaged in reuse, stories about reuse, or reuse ideas that have worked in your organization or at home.

Learn more and sign up at REUZEit.com.

Are Your Customers Motivated?

Beyond internal efforts, and to give extra weight to the value proposition of a circular business model, it certainly helps to attract customers that will care the most about your efforts. Data shows that a handful of audiences will embrace and even champion your CE efforts:

- Socially aware, ethical consumers
- Loyal brand followers
- Passionate green activists
- Millennials to a large extent, and Generation Z to an even larger one.

In their "2019 Retail and Sustainability Survey,"[175] CGS found that 68% of consumers think sustainability is important to their purchasing decisions, regardless of age or gender, and 35% of consumers would pay 25% more for sustainable products. A 2021 Futerra survey of over 1,000 US and UK consumers (summarized by Forbes[176]) found that 88% would like brands to help them be more environmentally friendly and ethical in their daily lives. In short, there awaits huge potential to use your brand power for good, with both loyal and even just slightly aware customers.

Importantly, looking ahead, NielsenIQ published a July 2024 report called "Spend Z"[177] that projects Gen Z to be the largest and wealthiest generation yet, with spending power estimated at $12T by 2030, set to overtake Boomer spend by 2029. They also found that Gen Zers report being concerned about environmental justice, and that they prefer to buy sustainable brands. Specifically, the report states that "products with multiple (two or more) sustainability attributes saw a 2.5 times higher sales lift than products that carried only a single sustainability claim." Another study, by First Insight, found that more than 62% of Gen Z consumers prefer to buy from a sustainable brand, with 73% of them willing to pay more for sustainable products.[178] If you're looking to engage future generations of consumers and get ahead of tapping into that spending power, circularity is a great path.

Does the opportunity seem appealing but the path to adoption seems long? Fear not— you don't have to close the loop on day one! You just need to get started. Initial steps toward circularity can make a huge difference with consumers. Launching a line of green products made from recycled or refurbished materials can signal that your brand is making strides toward sustainability. Including product descriptions and other educational materials about your efforts helps educate your customers and conveys authenticity and transparency at the same time. What's more, offering a warranty on refurbished items and/or developing a buyback system shows you're putting reuse into action. Finally, rewards programs and other incentives that encourage customers to join your sustainability efforts do double the work: they educate and strengthen retention.

[Side note: As a customer yourself, one way you can make a big difference is to buy from a circular economic platform. If you are going to buy used, buy from a reuse network rather than a traditional reseller. This goes for both things you buy for your home and your family as well as anything you might buy for your company—and if you buy equipment, don't forget about REUZEit!]

Education Is Power

Thinking even more broadly, and because the circular economy is still in its infancy, it's up to each of us to do what we can to educate others about the importance of the transition away from the status quo linear model, and how impactful it can be.

One great way to start a ripple effect is to train teachers, so that CE can be integrated into school and university curriculums, thus becoming an inherent part of the education of our future business leaders.

Companies can also run workshops for colleges and/or local community organizations that want to learn more about sustainability. Designing these educational events to be open and collaborative allows for productive information sharing and connection-making across industries and business functions.

Sustainability, Profitability, Solvability

REUZEit has been on the frontlines for more than a decade. When it comes to circularity, we know what works for big business, and it's not just about making things sustainable. You also need to reduce costs, generate profits, and provide solutions to unique challenges that come with the world of equipment. And when it comes to the technological equipment used to sustain our industries, continuing to buy new and throw out the old is hurting the environment more than many companies realize— especially when this equipment still holds value.

The good news? The upside opportunity is huge for both company and environment. As we've explored, keeping a piece of equipment in use—either by redeploying it internally or selling it on the used market— both benefits the bottom line and reduces GHG emissions about 23.1 times more than recycling.[179] With these numbers in mind, the solution to our growing, collective problem seems like a no-brainer.

There are avenues to help you manage your large volume of assets—and recover capital at the same time.

With REUZEit, you can finally gain a handle on your company's overwhelming surplus management. Put simply, our CE platform, centralized equipment market-place, and one-stop-shop services allow you to turn your company's surplus asset management strategy into a key sustainability driver, helping you streamline your processes with an end-to-end solution to asset visibility, sales reporting, logistics, and much more—all while reducing your carbon footprint.

The Future of Printers—and Beyond

It's no secret that humans are consuming more natural resources than we are putting back into the environment. Problems like pollution and supply chain short-ages are not going anywhere unless businesses are dedicated to the development of our new circular econ-omy. REUZEit found a significant amount of waste and inefficiency that occurs within large organizations and the use of machines that keep technology moving. We've been able to make a demonstrable change in our industry—and have had great business success in so doing, with $45M in assets under management in 2024 alone.

We need to rethink how we view the products that we buy. We can no longer afford to passively source materials, manufacture, use, and discard virtually everything we purchase.

Sometimes we think of the circular economy loop as Double Dutch jump roping. From the outside, it looks complex. An intricate dance based on interconnected rhythms, interdependent timing. A closed loop system. And just like learning a new skill like Double Dutch, it will take some time to move toward it and away from the linear economy. But there are steps that can be taken straight away.

Nowadays, Justin has a color laser, all-in-one printer at home. When it connects to WiFi, it works pretty well. The quality of the printing is much better than that printer from the '90s: high resolution, beautiful colors, long-lasting toner. There are days when Justin curses the machine because it inevitably requires a system update, or restarts on its own, or he resets the WiFi password in his house, and the machine won't do what he wants it to do. But the fact remains that this printer will become obsolete eventually because the entire machine can be brought down by a simple, impossible-to-replace plastic button malfunctioning. Or a software update that makes it no longer supportable by the manufacturer.

There is still a lot of work we need to do to change the way manufacturers make equipment. We need them to build things that last.

Now that the world has a platform for reuse—like ours, at REUZEit—it will be worth it for manufacturers to spend more on the shell of the printer: things like the frame, motors, trays, gears, and power supply. Instead of printers getting the eventual beat down they used to deserve, in the future, the shell of the machine could last your lifetime, or longer. It may require periodic upgrades, but we can make those core tech components easily available and ready to switch out. Transfer belts, scanner cameras, displays, and processors could be upgraded along with technological advancements for decades.

This is how the proverbial ball starts rolling.

Whether you work for a manufacturer, a logistics company, or a retailer: we're not asking you to embark on something on your own. Others have taken this path ahead of you. From smaller companies like Toast Brewing and Miuk Style to huge international names like Ikea and Patagonia, there are tons of case studies in this book—and countless more—to guide your way. We've been doing this for 15 years, and we're only just getting started. Our recent AI innovations have given us such a boost in efficiency that we're ready to scale. Justin recently moved his family to Southeast Asia to begin setting up partnerships and operations in that part of the world. All we see is potential. For us, for the environment, and for you.

The next time you walk into your workplace could be the beginning of a day like any other. But what could you do differently? What is one small step you could take toward circularity—toward shifting your organization, and the world, toward something that is truly sustainable?

Afterword

A Century Ahead: Envisioning a Fully Circular World

*T**hought exercises with Justin and Ryan Andrews: Imagine you are both in your 100s, the circular economy is well established, and the world has changed. Reflect on how all of it happened—how the shift came to be, and how we closed the loop.*

1. Reflecting on a Century of Transformation

Justin:

If someone had told us, back when we were bootstrapping surplus equipment in the 2010s, that circular marketplaces would one day *be* the world's main economy… I probably would've smiled politely and asked them to help us load a freezer into the truck.

Ryan:

Right? Back then, we were still arguing with people who thought "used" meant "useless." But here we are, in a world where waste hasn't just been minimized—it's been engineered out of existence.

Justin:

None of this could have happened through policy alone. It happened because economics flipped. Circularity became the most profitable and resilient way to operate—"anti-fragility" is a term coined by Nassim Nicholas Taleb, and it applies to most the business models of today who've had learned to thrive in economic adversity, not just survive it.

Ryan:

Circularity began as a movement for the environment, but what made it universal was profits and the holy grail of efficiency. Once society had platforms, data infrastructure, and modular design to match the ambition, circularity became the default—not the alternative.

The waste numbers we saw in the early 21st century were disgusting. Two billion tons of municipal waste per year. Less than 5% of the economy was considered circular back when we started. And let's not forget resource strain. Copper, lithium, rare earth elements—all of it was expected to run dry.

We were literally tearing apart the earth to make more of what we already had, just to let it sit, unused on warehouse shelves.

What changed wasn't just consumer behavior—it was the entire architecture of our systems. Value stopped being measured by how fast we could sell something new, and started being measured by how well we could keep it in play.

Justin:
And that's what this book was always about. Giving people a map—not just explaining the why, but showing the *how*. The circular future was always possible. We just had to build it.

2. The Circular Economy as the Core of Commerce

Ryan:
I think what would surprise people back in the 2020s most about the future we're living isn't that circularity *happened*—it's that it became the center of everything.

Justin:
Agreed. We used to treat it like a niche. Something you

added on circular practices after the "real" business was done. Today, circular marketplaces *are* the business. Most of the product flow moves through reuse networks, not traditional supply chains.

Ryan:

And it's not just secondhand stuff. New equipment still gets made, but it's designed with circularity as a prerequisite. Manufacturers assume that what they produce will be recovered, reconditioned and refurbished—multiple times over its lifespan.

Justin:

That shift required rethinking economics from the ground up. In a linear model, profit comes from turnover. But in a circular model, profit comes from durability, upgradeability, and the velocity of reuse.

Ryan:

It reminds me of when we started seeing equipment built with core/frame design. The frame—the outer shell, structural housing, high-value chassis—was built to last twenty, thirty, even fifty years. And the core? Swappable. Serviceable. Evolving with the tech.

Justin:

Exactly! That became the industrial standard. It started with high-value lab gear. Then hospital equipment, commercial appliances, even vehicles started following that modular logic. Why waste a perfectly engineered frame just because a processor or motor needs up-grading?

Ryan:

I remember a study—don't quote me on the number, but it showed something like 80% of the energy used to make a refrigerator was in the cabinet and insulation. The compressor was a tiny fraction. And yet, we were throwing out the whole thing for a slightly more efficient compressor or a software upgrade.

Justin:

Not anymore. In the fully circular economy, design starts with regeneration in mind. Every product is part asset, part inventory, part infrastructure. And that changes everything, from how insurance works, to how depreciation is calculated, to how we train engineers.

Ryan:

We went from a take-make-trash economy to a flow-

and-function economy. The product isn't a thing any-more—it's a node in a living system.

Justin:

And that system is now the engine of world commerce.

3. Evolution of Product Design: The Frame-and-Core Paradigm

Justin:

You know what really made the circular economy work? Design. Not just better logistics or regulations—*design thinking* at the product level. Once manufacturers embraced the frame-and-core paradigm, everything changed.

Ryan:

Totally. Before that shift, products were designed for obsolescence. Now, they're designed for evolution. Stronger, insulated, vacuum-sealed, corrosion-resistant shells. All you need to do to keep equipment at pace with fast-moving tech is replace the core: the sensors, processors, motors, or AI interfaces.

Justin:

It wasn't a new idea; it just finally matured. Consider a

lab freezer or industrial chiller: why toss the entire unit when all you need is a smarter compressor module or a new IoT board?

Ryan:

And it's not just about waste reduction. It's about lifecycle economics. By separating the slow and fast components of a product, we unlocked new value. Technicians can upgrade the core in hours, and companies can amortize the frame over decades.

Justin:

And here's the kicker: the secondhand market for cores became just as valuable as the one for frames. Because cores are refurbished, tested, and upgraded continuously. That created an entire secondary economy, filled with work, to manage all the components and close the loop.

Ryan:

It's basically what the Ellen MacArthur Foundation talked about years ago: *modularity as a path to material efficiency*. The future of manufacturing was always about designing for disassembly. We just didn't have the incentives—or the infrastructure—until the circular economy matured.

Justin:

And once modularity hit scale, it started shaping consumer expectations. People wanted products they could keep, customize, and maintain. The throwaway culture felt outdated, even stupid. Luxury brands were some of the first to adopt the model. High-end doesn't mean disposable anymore—it means, quite literally, *timeless*.

Ryan:

That's what circularity made possible: products that are not only better for the world, but better for people. Designed to last, to evolve, and to serve more than one owner, one use case, or one generation.

4. Human-Centered Labor in a Circular Economy

Justin:

If there's one thing people always got wrong about the circular economy, it's that they thought it was all about tech or AI robots taking everyone's jobs. That was true in new manufacturing, where robots could specialize on specific products and processes. But the deeper we got into the circular economy, the clearer it became: this is a labor story.

Ryan:

Absolutely. A circular economy doesn't just shift what we make—it reshapes *who* makes it, *who* repairs it, and *how* it keeps flowing. But with used equipment, there are too many variables for robots to handle. Too many unknowns and curve balls are required to talk to end users, identify excess equipment, uninstall and move it. Only a human has the dynamic capacity to inspect, combine or separate, refurbish, disassemble, repack, and recondition every component in the loop.

Justin:

In the linear economy, replacing equipment was almost always faster and "easier" than fixing it. But once circularity took over—once refurb became standard—it suddenly required a lot more hands, eyes, and care.

Ryan:

And that created jobs. Good, skilled jobs. Technicians, reconditioning specialists, warehouse managers, packaging engineers… Logistics went from being the forgotten backend to the engine of regeneration.

Justin:
It wasn't just about quantity either. It was about *meaning.* People took pride in what they were doing again. Instead of feeding a landfill, they were returning value to the system.

I remember seeing regional repair hubs in action—dozens of techs breaking down high-end scientific equipment, replacing parts, stress-testing systems. That kind of labor-intensive work would've been considered inefficient in the past. Now, it's the backbone of our economy.

5. Technological Advancements Supporting Circularity

Justin:
People always ask, "What made the circular economy scale?" And my answer is always vertical AI: *the tech finally caught up to the vision.* We used to dream about systems that could track every asset through its entire lifecycle. Now we have them. AI, IoT, digital twins, material passports—they made the invisible visible.

Ryan:

That's it. Once we could actually *see* where an asset had been, what condition it was in, how often it had been serviced—then reuse stopped being a gamble. It became a decision supported by data.

Justin:

And AI did more than just create listings or set market price. It became the co-pilot of circularity. Forecasting demand, suggesting ideal reuse pathways, and even flagging assets before they were at risk of being discarded.

I remember when "predictive maintenance" was a novelty. Now it's the baseline. Every piece of industrial equipment has a digital twin—an up-to-date model that tells technicians when to service it, what to replace, and what version the next core module needs to be.

Ryan:

And the material passport movement was huge. Once we had standardized tags that carried full component histories, the dataset knew every component, common fail points, how it was used, and even other uses for it in other industries.

Justin:

It's wild to think that in 2025, most companies didn't even have centralized surplus tracking. Fast-forward several decades, and now asset intelligence is baked into *everything*. Nothing gets lost. Nothing just sits.

Plus, smart manufacturing changed the game. Factories became flexible. They could pivot between producing new components and remanufacturing old ones. Machines didn't just build things, they rebuilt them.

Ryan:

What we learned was that AI and robots weren't there to replace human judgment—it was there to *amplify* it. The best circular systems were always AI-assisted but human-led. And that's the beauty of it. We didn't automate our way into sustainability. We collaborated our way into it. The future was built by technicians and drivers as much as it was by software.

6. Societal and Cultural Shifts

Ryan:

I think this is the part no one could've fully predicted: how deeply circularity would change *culture*.

Justin:

Totally. It started as an environmental and economic model, but it became something more. Once circular thinking took root, it started reshaping how people viewed ownership, responsibility, and even time.

Ownership was the first big shift. People stopped obsessing over *having* things and started focusing on *accessing* things. Equipment-as-a-service models exploded. Subscription platforms, shared tools, pooled infrastructure—it became normal to *use* without owning.

Ryan:

But it didn't feel like a sacrifice. In fact, people liked it. Fewer maintenance headaches. More flexibility. Products were always up to date, always working. You didn't need a garage full of gear—you needed access to a reliable circular network.

Justin:

That shift in mindset made other changes possible. For example, repair became aspirational. In the 20th century, people bragged about buying the newest thing. In the circular future, they brag about *how long something lasted*—and how many hands it passed through.

Looking back... it was huge. Circularity brought a new kind of status—longevity, care, and continuity. You didn't throw things away. You carried them forward. You cared for the next user without even knowing who they would be.

Ryan:
Education changed, too. Schools started teaching material literacy—how things are made, how they're deconstructed, how to design for regeneration. Kids grew up thinking about systems, not just products. Policy followed. Governments rewarded circular design and taxed waste; linear manufacturing wasn't outlawed, but it became expensive and inefficient, and the market simply evolved away from it.

Justin:
Culturally, it also reconnected us to *place*. Local economies thrived on circular hubs—repair shops, refurb centers, remanufacturing nodes. Circularity wasn't just for international enterprises—it was neighborhood-based.

That's such an underrated aspect, how the circular economy *localized* resilience. Every region had its own reuse ecosystems. Supply chains became more agile because they weren't single-threaded anymore—they were layered, distributed, circular.

Ryan:

People started seeing value *everywhere*. Not just in shiny new things, but in things that had history, use, and potential. That cultural shift, that sense of steward-ship over stuff, might've been the biggest transfor-mation of all.

7. Environmental and Economic Outcomes

Justin:

If we step back and look at the big picture—at what changed the world—it wasn't one innovation. It was the compounding effect of millions of small decisions that circularity made possible. The outcomes are stagger-ing. Resource extraction has plummeted. We used to dig, mine, and burn like there was no tomorrow. Now? Virgin resource demand has dropped by over 60% across the world. We're using what we already have—*smarter*.

Ryan:

And landfills? Mostly relics now. The few that remain are treated like resource banks, not dump sites. Old waste is mined, recovered, and reintroduced into the system. Waste isn't the end of a line anymore—it's the beginning of the next cycle.

And emissions. Let's not forget the role circularity played in reducing pollution. Reuse, refurbishment, modular upgrades, they all cut carbon across product lifecycles. It turned out that *using something twice* was better than building something "greener" from scratch.

Justin:

We also learned that circular economies are shock-proof. When supply chains hit turbulence, circular regions stayed afloat. Because they had inventory in motion, operations at business continued even if they couldn't buy new because of production halts.

Ryan:

That resilience became an economic advantage. Companies with closed-loop models, embracing anti-fragilely, outperformed traditional business in terms of stability and margin. Investors stopped chasing novelty and started betting on *durability*.

Justin:

And let's talk value. In the early days, reused equipment was considered second-tier—less valuable, less reliable. Now? A certified refurbished asset, with a verified service history and modular upgrade path, often sells for more than its linear equivalent.

It all comes back to information. Circular systems are transparent. Every component has a record, a rating, a path. That visibility reduces risk—and increases trust. And ultimately, that's the magic: circularity aligned profit with preservation. It proved that we don't need to exploit the environment to grow. We just need to rethink what growth means.

Ryan:

Exactly. The economy didn't shrink. It looped. And in doing so, it got stronger, cleaner, and more human.

8. The Journey Continues
Ryan:

So here we are, 100 years old, and we've seen the

circular economy evolve from an alternative idea into the backbone of modern civilization.

What we've learned is that sustainability isn't about sacrifice. It's about intelligence. It's about aligning the systems we build with the principles nature has always followed: maximizing use and minimizing waste.

Justin:
That mindset has unlocked more than environmental wins. It's given us economic resilience, social dignity, and technical mastery. But it's fragile, too. It only works if we keep iterating.

Ryan:
It only lasts if we teach it. If we pass it on. Reusing is not just a framework, it's a worldview. That regeneration isn't a niche, it's a responsibility. That long life is more valuable than fast growth.

Justin:
That's why we wrote this book back in 2025. Not just to document how circularity works today, but to inspire how it could work tomorrow. And the day after that.

Ryan:

So whether you're an engineer, a policymaker, a technician, a teacher—or someone just starting to ask questions—know this: the circular future isn't locked in.

Justin:

It's up to us. To build it. Maintain it. Improve it. And most of all—*believe in it.*

Ryan:

The story doesn't end here. It loops forward.

References

[1] Serio, Riccardo Gianluigi, Maria Michela Dickson, Diego Giuliani, and Giuseppe Espa. "Green Production as a Factor of Survival for Innovative Startups: Evidence from Italy." *Sustainability* 12, no. 22 (2020): 9464. November 13, 2020, accessed June 21, 2025. https://iris.unitn.it/retrieve/handle/11572/281486/384992/sustainability-12-09464.pdf.

[2] Randi Kronthal-Sacco and Tensie Whelan, *2022 Sustainable Market Share Index™: Research on IRI Purchasing Data of U.S. Consumer Packaged Goods* (New York: NYU Stern Center for Sustainable Business, April 2023), 1, https://www.stern.nyu.edu/sites/default/files/2023-04/FINAL%202022%20CSB%20Report%20for%20website.pdf.

[3] Sherry Frey, Jordan Bar Am, Vinit Doshi, Anandi Malik, and Steve Noble, "Consumers Care About Sustainability—and Back It Up with Their Wallets," McKinsey & Company, February 16, 2023, accessed June 21, 2025, https://www.mckinsey.com/industries/consumer-packaged-goods/our-insights/consumers-care-about-sustainability-and-back-it-up-with-their-wallets.

[4] PwC, "Consumers Respond to Waves of Disruption: June 2022 Global Consumer Insights Pulse Survey," June 2022, https://www.pwc.com/gx/en/consumer-markets/consumers-respond-to-waves-of-disruption/gcis-report-june-2022.pdf.

[5] Thomson Reuters Institute, "The 2023 State of Corporate ESG: How Companies Are Embracing ESG for Resilience and Growth," Thomson Reuters, November 7, 2023, accessed June 21, 2025, https://www.thomsonreuters.com/en-us/posts/esg/state-of-corporate-esg-report-2023/.

[6] Paul van den Brande, "Reputational Damage," LinkedIn, June 19, 2024, accessed June 21, 2025, https://www.linkedin.com/pulse/reputational-damage-paul-van-den-brande-rdz0e/.

[7] Brian Kennedy and Alec Tyson, Pew Research Center, "How Americans View Climate Change and Policies to Address the Issue," Pew Research Center, December 9, 2024, https://www.pewresearch.org/science/2024/12/09/how-americans-view-climate-change-and-policies-to-address-the-issue/#majority-of-

americans-expect-to-make-sacrifices-due-to-climate-change-but-rela-tively-few-expect-them-to-be-major.

[8] Ron Gonen, *The Waste-Free World: How the Circular Economy Will Take Less, Make More, and Save the Planet* (New York: Portfolio/Penguin, 2021), xvii.

[9] Ellen MacArthur Foundation, "The Circular Economy in Detail," Ellen MacArthur Foundation, September 15, 2019, accessed June 21, 2025, https://www.ellenmacarthurfoundation.org/the-circular-economy-in-detail-deep-dive.

[10] Ellen MacArthur Foundation, *Completing the picture: How the circular economy tackles climate change* (2019).

[11] Ellen MacArthur Foundation, "Production of Nylon Yarn from Waste Materials," October 5, 2021, accessed June 21, 2025, https://www.ellenmacarthurfoundation.org/circular-examples/production-of-nylon-yarn-from-waste-materials.

[12] McKinsey & Company, "Growth Within: A Circular Economy Vision for a Competitive Europe," June 1, 2015, accessed September 26, 2024, https://www.mckinsey.com/capabilities/sustainability/our-insights/growth-within-a-circular-economy-vision-for-a-competitive-europe#/.

[13] Ellen MacArthur Foundation, "Breathing New Life into Vacant Commercial Buildings: Novaxia," August 12, 2024, accessed June 21, 2025, https://www.ellenmacarthurfoundation.org/breathing-new-life-into-vacant-commercial-buildings-novaxia.

[14] Merriam-Webster, s.v. "innovation," accessed June 21, 2025, https://www.merriam-webster.com/dictionary/innovation.

[15] Gonen, *The Waste-Free World*, 35–6.

[16] Circle Economy, "CGR 2021," accessed October 4, 2024, https://www.circularity-gap.world/2021, under "Interventions Vortex."

[17] United Nations Environment Programme and International Solid Waste Association, "Global Waste Management Outlook 2024," UNEP, March 2024, https://www.unep.org/resources/global-waste-management-outlook-2024.

[18] IKEA, "A Circular IKEA – Making the Things We Love Last Longer," accessed October 8, 2024, https://www.ikea.com/us/en/this-is-ikea/sustainable-everyday/a-circular-ikea-making-the-things-we-love-last-longer-pub9750dd90.

[19] Hanna Andersson, "Hanna-Me-Downs," accessed January 22, 2025, https://preloved.hannaandersson.com/.

[20] William McDonough and Michael Braungart, *Cradle to Cradle: Remaking the Way We Make Things* (New York: North Point Press, 2002), 20–21.

[21] World Bank, "What Is the Blue Economy?" infographic, June 6, 2017, https://www.worldbank.org/en/news/info-graphic/2017/06/06/blue-economy.

[22] ING Research. "Opportunity and Disruption: How Circular Thinking Could Change US Business Models." ING, February 5, 2019, accessed October 2, 2024. https://www.ing.com/MediaEditPage/Opportunity-and-disruption-How-circular-thinking-could-change-US-business-models.htm.

[23] Samuel J.G. Cooper, Jannik Giesekam, Geoffrey P. Hammond, Jonathan B. Norman, Anne Owen, John G. Rogers, Kate Scott, Thermodynamic insights and assessment of the 'circular economy', Journal of Cleaner Production, Volume 162, 2017, Pages 1356-1367, ISSN 0959-6526, https://doi.org/10.1016/j.jclepro.2017.06.169. (https://www.sciencedirect.com/science/article/pii/S0959652617313392, accessed October 2, 2024)

[24] Global Environment Facility, "Land Degradation," accessed October 2, 2024, https://www.thegef.org/what-we-do/topics/land-degradation.

[25] Ellen MacArthur Foundation, "A Regenerative Supply Chain Alternative for Palm Oil," April 22, 2022, accessed October 2, 2024, https://www.ellenmacarthurfoundation.org/circular-examples/a-regenerative-supply-chain-alternative-for-palm-oil.

[26] Gonen, *The Waste-Free World*, 62.

[27] E. F. Schumacher, *Small Is Beautiful* (London: Blond & Briggs, 1973), 15.

28 World Economic Forum, "Why the Circular Economy Is the Business Opportunity of Our Time," May 24, 2022, https://www.weforum.org/agenda/2022/05/why-the-circular-economy-is-the-business-opportunity-of-our-time/.

29 Platform for Accelerating the Circular Economy (PACE) and Boston Consulting Group, "The Circular Economy Landscape: Metrics for Measuring Progress," white paper, October 2020, accessed October 2, 2024, https://pacecircular.org/sites/default/files/2021-03/5faa4d272e1a82a1d9126772_20201029%20-%20BCG%20Metrics%20-%20White%20Papers%20-%20The%20Landscape%20-%20210_x_297_mm%20-%20bleed_3_mm.pdf.

30 Ellen MacArthur Foundation, *A new textiles economy: Redesigning fashion's future* (2017).

31 Philip J. Landrigan et al., "The Lancet Commission on pollution and health," *The Lancet* 391, no. 10119 (February 3, 2018): 462–512, https://www.thelancet.com/commissions/pollution-and-health.

32 United Nations Environment Programme, "UNEP and Biodiversity," briefing note, September 2020, accessed October 9, 2024, https://www.unep.org/unep-and-biodiversity.

33 Simon Reddy and Winnie Lau, "Breaking the Plastic Wave: Top Findings for Preventing Plastic Pollution," The Pew Charitable Trusts, July 23, 2020, accessed October 9, 2024, https://www.pewtrusts.org/en/research-and-analysis/articles/2020/07/23/breaking-the-plastic-wave-top-findings.

34 United Nations Environment Programme, "Valuing Plastic: The Business Case for Measuring, Managing and Disclosing Plastic Use in the Consumer Goods Industry," 2014, 12, accessed October 9, 2024, https://wedocs.unep.org/bitstream/handle/20.500.11822/9238/-Valuing%20plastic%3a%20the%20business%20case%20for%20measuring%2c%20managing%20and%20disclosing%20plastic%20use%20in%20the%20consumer%20goods%20industry-2014Valuing%20plasticsF.pdf?sequence=8&is.

35 ORCID, "Harold Krikke," accessed June 21, 2025, https://orcid.org/0000-0002-9010-9645.

[36] Krikke, Harold, Nestor Coronado Palma, Jonah Shell, and Justin Andrews. "Circular Economic Surplus Asset Management: A Game Changer in Life Sciences." *IEEE Engineering Management Review* 50, no. 2 (June 2022): 117–26. https://doi.org/10.1109/EMR.2022.3174634.

[37] Krikke et al., "Circular Economic Surplus Asset Management," 117–26.

[38] Ellen MacArthur Foundation, "Upstream Innovation: Overview," October 21, 2022, accessed October 10, 2024, https://www.ellenmacarthurfoundation.org/upstream-innovation/overview.

[39] Susi Wallner, StartUs Insights, "Top 8 Circular Economy Trends & Innovations in 2021," June 17, 2021, accessed October 10, 2024, https://www.startus-insights.com/innovators-guide/top-8-circular-economy-trends-innovations-in-2021/.

[40] European Commission, "Single-Use Plastics," accessed October 11, 2024, https://environment.ec.europa.eu/topics/plastics/single-use-plastics_en.

[41] U.S. Department of Energy, "Biden Administration Implements New Cost-Saving Energy Efficiency Standards for Light Bulbs," Energy.gov, accessed October 11, 2024, https://www.energy.gov/articles/biden-administration-implements-new-cost-saving-energy-efficiency-standards-light-bulbs.

[42] Lexie Pelchen, "US Incandescent Light Bulb Ban," Forbes, updated July 1, 2024, accessed October 11, 2024, https://www.forbes.com/home-improvement/electrical/us-incandescent-light-bulb-ban/.

[43] National Conference of State Legislatures, "Right to Repair 2023 Legislation," NCSL.org, updated November 1, 2023, accessed October 11, 2024, https://www.ncsl.org/technology-and-communication/right-to-repair-2023-legislation.

[44] Patagonia, "Start Repair," Patagonia.com, accessed October 11, 2024, https://www.patagonia.com/start-repair/.

[45] Amazon, "Certified Refurbished," Amazon.com, accessed January 23, 2025, https://www.amazon.com/Certified-Refurbished/b?node=12653393011.

[46] Karino Ayako, "Beer to Paper: Upcycling Malt Lees into Craft Beer Paper," Zenbird, July 22, 2021, accessed October 11, 2024, https://zenbird.media/beer-to-paper-upcycling-malt-lees-into-craft-beer-paper/.

[47] Zenbird Media, "Introducing World's First Incense Made from Upcycled Cacao Husks," September 22, 2024, accessed October 11, 2024, https://zenbird.media/introducing-worlds-first-incense-made-from-upcycled-cacao-husks/.

[48] "Circular Economy Business Models," Edison Group, March 24, 2022, accessed October 11, 2024, https://www.edisongroup.com/thematic/circular-economy-business-models/.

[49] "Lansing City, Michigan," U.S. Census Bureau, accessed November 1, 2024, https://data.census.gov/profile/Lansing_city,_Michigan?g=160XX00US2646000#housing.

[50] Jennifer Melroy, "US National Parks Ranked by Size," National Park Obsessed, published August 7, 2020, updated September 16, 2022, accessed November 1, 2024, https://nationalparkobsessed.com/national-parks-by-size/#1-us-national-parks-ranked-by-size.

[51] Alex Wilhelm, "Turo Car Rental IPO Profitable Growth," TechCrunch, March 8, 2024, accessed October 11, 2024, https://techcrunch.com/2024/03/08/turo-car-rental-ipo-profitable-growth/.

[52] Matt Weller, "Turo IPO: Everything You Need to Know About Turo," Forex.com, January 18, 2024, accessed October 11, 2024, https://www.forex.com/en-us/news-and-analysis/turo-ipo-everything-you-need-to-know-about-turo/.

[53] Circular Innovation, "Circular Business Models: Product Life Extension," accessed October 18, 2024, https://circularinnovation.ca/circular-business-models-product-life-extension/.

[54] Y. Leterrier, "Polymer Matrix Composites," in *Comprehensive Composite Materials*, eds. Anthony Kelly and Carl Zweben (Oxford: Elsevier, 2000), 1073–1102, sec. 2.33.3.2.2, accessed October 18, 2024, https://www.sciencedirect.com/reference-work/9780080429939/comprehensive-composite-materials.

[55] "Disseny per al Reciclatge" (Design for Recycling), accessed October 18, 2024, http://www.arc-cat.net/en/publica-cions/pdf/ccr/diss_reci.pdf.

[56] Product Stewardship Institute, "Tires," accessed October 18, 2024, https://productstewardship.us/products/tires/.

[57] Product Stewardship Institute, "Tires."

[58] The Six Principles of Regenerative Farming: Why Are They Important?," AgriCapture CO₂, accessed October 18, 2024, https://agri-captureco2.eu/the-six-principles-of-regenerative-farming-why-are-they-important/.

[59] "Regenerative Agriculture 101," Natural Resources Defense Council, November 29, 2021, accessed October 18, 2024, https://www.nrdc.org/stories/regenerative-agriculture-101.

[60] "Keeping Grasslands Healthy in Northern Tanzania," The Nature Conservancy, January 26, 2020, accessed October 18, 2024, https://www.nature.org/en-us/about-us/where-we-work/africa/stories-in-africa/healthy-grasslands-tanzania/.

[61] Jill Barth, "Beyond Organic: The Winemakers Leading a Sustainable Revolution," Wine Enthusiast, updated June 27, 2023, accessed October 18, 2024, https://www.wineenthusiast.com/culture/wine/regenera-tive-agriculture-wine-organic/.

[62] "Regenerative Agriculture 101," NRDC (principles section).

[63] "What Is Regenerative Agriculture?," Rainforest Alliance, updated October 7, 2024, accessed October 18, 2024, https://www.rainforest-al-liance.org/insights/what-is-regenerative-agriculture/.

[64] Janine M. Benyus, *Biomimicry: Innovation Inspired by Nature* (London: HarperCollins, 2009).

[65] Gertie Goddard, "Biomimetic Design: 10 Examples of Nature Inspiring Technology," BBC Science Focus, accessed October 18, 2024, https://www.sciencefocus.com/future-technology/biomimetic-design-10-examples-of-nature-inspiring-technology.

[66] Henry Beniard, "Circular Economy Strategy," Twice Commerce (blog), June 13, 2023, updated November 17, 2023, accessed June 4, 2025, https://www.twicecommerce.com/blog/circular-economy-strategy.

[67] Nancy M.P. Bocken and Thijs H.J. Geradts, "Designing Your Circular Business Model," Stanford Social Innovation Review, Spring 2022, accessed June 4, 2025, https://ssir.org/articles/entry/designing_your_circular_business_model.

[68] Peter Lacy, Jessica Long, and Wesley Spindler, *The Circular Economy Handbook: Realizing the Circular Advantage* (Accenture, 2020), 133.

[69] Priliantina Bebasari, "The Role of Women in Upcycling Initiatives in Jakarta, Indonesia: A Case for the Circular Economy in a Developing Country," in *The Circular Economy in the Global South*, eds. Patrick Schröder, Manisha Anantharaman, Kartika Anggraeni, and Timothy J. Foxon (Routledge, 2019), 75.

[70] "CE-SAM White Paper," Reuzeit, accessed October 29, 2024, https://www.reuzeit.com/white-paper/ce-sam.

[71] Internal Revenue Service, "How to Depreciate Property," Publication 946 (2023), accessed December 30, 2024, https://www.irs.gov/pub/irs-pdf/p946.pdf.

[72] ToolSense, "Asset Lifespan: How to Calculate and Extend the Useful Life of Assets," accessed October 30, 2024, https://toolsense.io/business/asset-lifespan-how-to-calculate-and-extend-the-useful-life-of-assets/.

[73] "International Organization for Standardization," Wikipedia, accessed January 28, 2025, https://en.wikipedia.org/wiki/International_Organization_for_Standardization.

[74] International Organization for Standardization, "ISO 59004:2024 Circular economy — Vocabulary, principles and guidance for implementation," May 2024, accessed December 30, 2024, https://www.iso.org/standard/80648.html.

[75] International Organization for Standardization, "ISO 59010:2024 Circular economy — Guidance on the transition of business models and value networks," May 2024, accessed December 30, 2024, https://www.iso.org/standard/80649.html.

[76] International Organization for Standardization, "ISO 59020:2024 Circular economy — Measuring and assessing circularity performance," May 2024, accessed December 30, 2024, https://www.iso.org/standard/80650.html.

[77] International Organization for Standardization, "ISO 59014:2024 Environmental management and circular economy — Sustainability and traceability of the recovery of secondary materials — Principles, requirements and guidance," October 2024, accessed December 30, 2024, https://www.iso.org/standard/80694.html.

[78] International Organization for Standardization, "ISO 59040:2025 Circular economy — Product circularity data sheet," February 2025, accessed June 21, 2025, https://www.iso.org/standard/82339.html.

[79] International Organization for Standardization (ISO), *ISO/TS 55010:2024—Guidance on the Alignment of Financial and Non-Financial Functions in Asset Management*, published July 2024, accessed June 4, 2025, https://www.iso.org/standard/84051.html.

[80] European Commission, "Decommissioning of Nuclear Facilities," accessed November 4, 2024, https://energy.ec.europa.eu/topics/nuclear-energy/decommissioning-nuclear-facilities_en.

[81] Ellen MacArthur Foundation, "Food and the Circular Economy Deep Dive," September 15, 2019, accessed June 4, 2025, https://www.ellenmacarthurfoundation.org/food-and-the-circular-economy-deep-dive.

[82] Ellen MacArthur Foundation, "Food and the Circular Economy Deep Dive."

[83] Ellen MacArthur Foundation, "Food and the Circular Economy Deep Dive."

[84] Chris Voloschuk, "Aquapak Study Says Shift from Plastic to Paper Packaging Moving Slowly," Recycling Today, May 14, 2024, accessed November 4, 2024, https://www.recyclingtoday.com/news/aquapak-study-says-shift-from-plastic-to-paper-packaging-moving-slowly/.

[85] Miloš Rajković, Dušanka Popović Minić, Danijel Milinčić, and Milena Zdravković, "Circular Economy in Food Industry," *Zastita materijala* 61 (2020): 229–250, https://doi.org/10.5937/zasmat2003229R6.

[86] Trader Joe's, "Neighborhood Shares," accessed November 13, 2024, https://www.traderjoes.com/home/neighborhood-shares.

[87] Cary Funk and Brian Kennedy, "Americans' Views About and Consumption of Organic Foods," Pew Research Center, December 1, 2016, accessed November 13, 2024, https://www.pewresearch.org/science/2016/12/01/americans-views-about-and-consumption-of-organic-foods/.

[88] "Community Gardens Increase Food Security and Community Well-Being," Land-Grant Impacts, accessed November 13, 2024, https://landgrantimpacts.org/community-gardens-increase-food-security-and-community-well-being/.

[89] National Land-Grant Impacts Database, "Feeding Homeless and Teaching Community to Grow," University of Georgia Cooperative Extension, 2023, accessed November 13, 2024, https://nidb.landgrantimpacts.org/impacts/show/6641.

[90] Francesca Vergani, "Higher Education Institutions as a Microcosm of the Circular Economy," *Journal of Cleaner Production* 435 (2024): 140592, https://doi.org/10.1016/j.jclepro.2024.140592.

[91] UNESCO, "UNESCO-Japan Prize on Education for Sustainable Development Awards Projects from Guatemala, Japan and Zimbabwe," published November 2, 2023, updated September 14, 2024, accessed November 13, 2024, https://www.unesco.org/en/articles/unesco-japan-prize-education-sustainable-development-awards-projects-guatemala-japan-and-zimbabwe.

[92] UNESCO Institute for Lifelong Learning, "Education for Sustainable Development: Progress and Challenges," 2021, accessed November 13, 2024, https://unesdoc.unesco.org/ark:/48223/pf0000379535.locale=en.

[93] Deloitte, "Sustainability in Retail," February 2023, accessed November 14, 2024, https://www.deloitte.com/global/en/Industries/consumer/analysis/sustainability-in-retail.html.

[94] Deloitte, "Sustainability in Retail."

[95] H&M Group, "Collect, Recirculate, Recycle," accessed November 14, 2024, https://hmgroup.com/sustainability/circularity-and-climate/recycling/.

[96] Asda, "We've Opened the UK's Largest Refill Store as Part of Our Drive to Reduce Plastic," October 7, 2021, accessed November 15, 2024, https://corporate.asda.com/20211007/weve-opened-the-uks-largest-refill-store-as-part-of-our-drive-to-reduce-plastic

[97] Statista, "Electronic Waste Worldwide," May 6, 2024, accessed November 15, 2024, https://www.statista.com/topics/3409/electronic-waste-worldwide/.

[98] Apple, "Environment," accessed November 15, 2024, https://www.apple.com/environment/.

[99] Kahupi, Inamutila, Natalia Yakovleva, Okechukwu Okorie, and Clyde Eiríkur Hull. "Implementation of Circular Economy in a Developing Economy's Mining Industry Using Institutional Theory: The Case of Namibia." *Journal of Environmental Management* 368, no. 122145 (September 2024). https://doi.org/10.1016/j.jenvman.2024.122145.

[100] International Copper Association, *Copper—The Pathway to Net Zero: Summary* (March 2023), accessed November 15, 2024, https://internationalcopper.org/wp-content/uploads/2023/03/ICA-GlobalDecarb-Summary-A4-202302-R3.pdf.

[101] International Copper Association, *Copper Recycling: The Key to Circular Economy* (January 2022), accessed November 15, 2024, https://internationalcopper.org/wp-content/uploads/2022/02/ICA-RecyclingBrief-202201-A4-R2.pdf.

[102] Louise Assem, "The Role of Mining in the Circular Economy," International Copper Association, published October 20, 2023, accessed November 15, 2024, https://internationalcopper.org/resource/the-role-of-mining-in-the-circular-economy/.

[103] Matthew Parizot, "Taking a Circular Approach to Mining Operations," *CIM Magazine*, January 19, 2022, accessed November 15, 2024, https://magazine.cim.org/en/news/2022/taking-a-circular-approach-to-mining-operations-en/.

[104] International Civil Aviation Organization, "Circular Economy," accessed November 15, 2024, https://www.icao.int/environmental-protection/Pages/CircularEconomy.aspx.

[105] Air Transport Action Group, "Circular Economy," accessed June 10, 2025, https://aviationbenefits.org/environmental-efficiency/circular-economy/.

[106] Okumus, Dogancan, Sefer A. Gunbeyaz, Rafet E. Kurt, and Osman Turan. 2024. "An Approach to Advance Circular Practices in the Maritime Industry through a Database as a Bridging Solution" *Sustainability* 16, no. 1: 453. https://doi.org/10.3390/su16010453.

[107] Kelsey Miller, "The Triple Bottom Line: What It Is & Why It's Important," Harvard Business School Online, updated August 16, 2023, accessed June 5, 2025, https://online.hbs.edu/blog/post/what-is-the-triple-bottom-line.
[108] Miller, "The Triple Bottom Line."

[109] Karl Haller, Mary Wallace, Jane Cheung, and Sachin Gupta, *Consumers Want It All: Hybrid Shopping, Sustainability, and Purpose-Driven Brands* (IBM Institute for Business Value, 2022), accessed November 21, 2024, https://www.ibm.com/downloads/cas/YZYLMLEV.

[110] Simon-Kucher & Partners, *Global Sustainability Study 2021*, accessed November 21, 2024, https://www.simon-kucher.com/sites/default/files/studies/Simon-Kucher_Global_Sustainability_Study_2021.pdf.

[111] Minnesota Pollution Control Agency, *Reducing Waste in the Workplace*, accessed June 5, 2024, https://www.pca.state.mn.us/sites/default/files/w-hhw1-14.pdf.

[112] Hayley Peterson, "Why Etsy Got Rid of Individual Employee Trash Cans," Business Insider, September 9, 2014, accessed November 21, 2024, https://www.businessinsider.com/why-etsy-got-rid-of-individual-employee-trash-cans-2014-9.

[113] ANEW, "History & Vision," accessed November 22, 2024, https://anewfound.org/about-anew/history-vision/.

[114] Heather Goetsch and Michael Deru, *Operational Emissions Accounting for Commercial Buildings* (Golden, CO: National Renewable Energy Laboratory, July 2022), NREL/TP-5500-81670, accessed November 21, 2024, https://www.nrel.gov/docs/fy22osti/81670.pdf.

[115] Mohamed Elhassan, "Circular Economy in the Management of End-of-Life Solar Panels," *Ecology & Conservation of Environment* 6, no. 1 (2023): 1–7, accessed November 21, 2024, https://medwinpublishers.com/EOIJ/EOIJ16000166.pdf.

[116] Jeanne C. Meister, "The #1 Office Perk? Natural Light," Harvard Business Review, September 3, 2018, accessed November 21, 2024, https://hbr.org/2018/09/the-1-office-perk-natural-light.

[117] U.S. Department of Energy, "LED Lighting," accessed November 21, 2024, https://www.energy.gov/energysaver/led-lighting.

[118] International Renewable Energy Agency (IRENA), *Renewable Power Generation Costs in 2020* (June 2021), accessed November 21, 2024, https://www.irena.org/publications/2021/Jun/Renewable-Power-Costs-in-2020.

[119] National Grid, "What Are Scope 1, 2 and 3 Carbon Emissions?," last updated July 1, 2024, accessed November 22, 2024, https://www.nationalgrid.com/stories/energy-explained/what-are-scope-1-2-3-carbon-emissions.

[120] International Air Transport Association, "IATA Economics' Chart of the Week: Parked Aircraft Returning to Service Support Global Recovery," March 31, 2023, accessed December 30, 2024, https://www.iata.org/en/iata-repository/publications/economic-reports/parked-aircraft-returning-to-service-support-global-recovery/.

[121] Boeing, "Boeing Forecasts Demand for Nearly 44,000 New Airplanes Through 2043 as Air Travel Surpasses Pre-Pandemic Levels," July 20, 2024, accessed December 30, 2024, https://boeing.mediaroom.com/2024-07-19-Boeing-Forecasts-Demand-for-Nearly-44,000-New-Airplanes-Through-2043-as-Air-Travel-Surpasses-Pre-Pandemic-Levels.

[122] Seventh Generation, "More Than a Name," January 1, 2022, accessed November 22, 2024, https://www.seventhgeneration.com/blog/more-than-a-name.

[123] Beth Kowitt, "Unilever Is Buying Seventh Generation," Fortune, September 19, 2016, accessed November 22, 2024, https://fortune.com/2016/09/19/unilever-buying-seventh-generation/.

[124] Alden Zecha, "What the Unilever Seventh Generation Acquisition Means for Social Ventures," CASE at Duke, October 6, 2016, accessed November 22, 2024, https://centers.fuqua.duke.edu/case/2016/10/unilever-seventh-generation/.

[125] Merriam-Webster, s.v. "artificial intelligence," definition 2, accessed December 4, 2024, https://www.merriam-webster.com/dictionary/artificial%20intelligence

[126] Thor Olavsrud, "5 Famous Analytics and AI Disasters," CIO, updated June 9, 2025, https://www.cio.com/article/190888/5-famous-analytics-and-ai-disasters.html.

[127] Reddit, "r/aifails," last accessed June 23, 2025, https://www.reddit.com/r/aifails/.

[128] United Nations Environment Programme, "AI Has an Environmental Problem. Here's What the World Can Do About That," September 21, 2024, accessed February 5, 2025, https://www.unep.org/news-and-stories/story/ai-has-environmental-problem-heres-what-world-can-do-about; David M. Hart and Lauren E. Riggs, "The Uneven Distribution of AI's Environmental Impacts," Harvard Business Review, July 2024, accessed February 5, 2025, https://hbr.org/2024/07/the-uneven-distribution-of-ais-environmental-impacts; Yuan Yao, "Can We Mitigate AI's Environmental Impacts?" Yale School of the Environment, October 10, 2024, accessed February 5, 2025, https://environment.yale.edu/news/article/can-we-mitigate-ais-environmental-impacts.

[129] Ellen MacArthur Foundation, *Artificial intelligence and the circular economy: AI as a tool to accelerate the transition* (2019).

[130] Artificial Intelligence and Robotization in Waste Sorting Centers as an Alternative to Brown Bins," BiogasWorld, September 26, 2017, accessed June 9, 2025, https://biogasworld.com/news/artificial-intelligence-robotization-waste-sorting-centers-alternative-brown-bins/;

"How ZenRobotics Sorting Technology Is Helping C&D Recyclers Pick Up to 12,000 Items per Hour," Recycling Product News, February 9, 2022, accessed June 9, 2025, https://www.recyclingproduct-news.com/article/38118/how-zenrobotics-sorting-technology-is-help-ing-candd-recyclers-pick-up-to-12000-items-per-hour.

[131] Ellen MacArthur Foundation, *Artificial Intelligence and the Circular Economy*, 5.

[132] Michael Leitl, "Circular Economy Potential: 6 AI Use Cases for Sustainability," Indeed Innovation, September 11, 2023, accessed December 4, 2024, https://www.indeed-innovation.com/the-mensch/ai-use-cases-circular-economy/.

[133] Andrei Klubnikin (Innovation Analyst), Vitali Likhadzed (ITRex CEO), Kirill Stashevsky (ITRex CTO), and Nadejda Alkhaldi (Innovation Analyst), "How to Implement AI in Business: The Definitive Guide," ITRex Group, published February 13, 2025, accessed December 5, 2024, https://itrexgroup.com/blog/how-to-implement-ai-in-busi-ness/.

[134] Walter R. Stahel, *The Circular Economy: A User's Guide* (Routledge, 2019), 85.

[135] Stats in the Belgium section were all pulled from United Nations Environment Programme, "Belgium on Its Way Towards a Circular Economy," February 25, 2020, accessed December 5, 2024, https://www.unep.org/news-and-stories/story/belgium-its-way-towards-circular-economy.

[136] France, Law No. 2015-992 on Energy Transition for Green Growth (Energy Transition Law), accessed June 9, 2025, https://climate-laws.org/document/law-no-2015-992-on-energy-transition-for-green-growth-energy-transition-law_aea3.

[137] "Circular Economy Roadmap: 50 Measures for a 100% Circular Economy," Circular Economy Roadmap France, April 2018, accessed June 9, 2025, https://circulareconomy.europa.eu/platform/en/strate-gies/circular-economy-roadmap-france-50-measures-100-circular-econ-omy

[138] France, *LOI n° 2020-105 du 10 février 2020 relative à la lutte contre le gaspillage et à l'économie circulaire*, accessed June 9, 2025, https://www.legifrance.gouv.fr/jorf/id/JORFTEXT00004155375 9/.

[139] Ellen MacArthur Foundation, "France's Anti-waste and Circular Economy Law: Eliminating Waste and Promoting Social Inclusion," case study, August 2021, accessed June 9, 2025, https://circulareconomy.europa.eu/platform/sites/default/files/case_studies_-_french_anti_waste_law_aug21.pdf.pdf.

[140] Organisation for Economic Co-operation and Development, *Effective Carbon Rates 2021*, May 5, 2021, accessed June 9, 2025, https://www.oecd.org/en/publications/effective-carbon-rates-2021_0e8e24f5-en.html

[141] Krikke et al., "Circular Economic Surplus Asset Management," 117–26.

[142] Circle Economy, *CGR 2021*.

[143] INNOWEEE, "Italian Circular Economy Report 2020," accessed June 10, 2025, https://www.innoweee.eu/CircularEconomy2020.

[144] Italian Ministry of Economy and Finance, "Less Taxes and More Focus on the Environment and Welfare: The 2020 Budget Law," March 6, 2020, accessed June 10, 2025, https://www.mef.gov.it/en/focus/Less-taxes-and-more-focus-on-the-environment-and-welfare-the-2020-Budget-Law/.

[145] European Commission, "The European Green Deal," strategy document, accessed May 2, 2025, https://commission.europa.eu/strategy-and-policy/priorities-2019-2024/european-green-deal_en.

[146] The 3rd Industrial Revolution Consulting Group LLC and Luxembourg Working Groups, *The 3rd Industrial Revolution Strategy Study for the Grand Duchy of Luxembourg: Thematic Summary* (Luxembourg: TIR Consulting Group LLC, 2016), accessed December 5, 2024, https://www.troisiemerevolutionindustrielle.lu/wp-content/uploads/2016/11/TIR-Strategy-Study_Short.pdf.

[147] Government of the Netherlands, *A Circular Economy in the Netherlands by 2050* (2016), accessed December 5, 2024, https://circulareconomy.europa.eu/platform/sites/default/files/17037circulaireeconomie_en.pdf.

[148] Government of the Netherlands, *A Circular Economy in the Netherlands by 2050*, 6.

[149] Government of Portugal, *Leading the Transition: Action Plan for Circular Economy in Portugal 2017–2020* (2017), accessed June 10, 2025, https://circulareconomy.europa.eu/platform/sites/default/files/strategy_-_portuguese_action_plan_paec_en_version_3.pdf.

[150] Ministerio para la Transición Ecológica y el Reto Demográfico, *Estrategia Española de Economía Circular*, accessed June 10, 2025, https://www.miteco.gob.es/es/calidad-y-evaluacion-ambiental/temas/economia-circular/estrategia.html.

[151] Construcía, "Moving towards Circular Economy and Zero Waste," accessed June 23, 2025, https://www.construcia.com/en/noticias/moving-towards-circular-economy-and-zero-waste/.

[152] UK Government, "UK to Establish World's First UN-Backed Centre for Circular Economy Research," February 27, 2024, accessed December 6, 2024, https://www.gov.uk/government/news/uk-to-establish-worlds-first-un-backed-centre-for-circular-economy-research.

[153] Circle Economy, "United Kingdom Circularity Gap Report," accessed December 6, 2024, https://www.circularity-gap.world/united-kingdom.

[154] Center for Sustainable Systems, University of Michigan, "U.S. Material Use Factsheet," accessed June 10, 2025, https://css.umich.edu/publications/factsheets/material-resources/us-material-use-factsheet.

[155] Center for Sustainable Systems, "U.S. Material Use Factsheet."

[156] Center for Sustainable Systems, "U.S. Material Use Factsheet."

[157] Circular CoLab, "U.S. Circular Economy Report," October 2018, accessed June 10, 2025, https://www.circularcolab.org/us-circular-economy-report.

[158] Kweku Attafuah-Wadee and Johanna Tilkanen, "Accelerating the Circular Economy Transition in Africa: Policy Challenges and Opportunities," Circle Economy Earth, published November 2, 2020, accessed June 10, 2025, https://circulareconomy.earth/publications/accelerating-the-circular-economy-transition-in-africa-policy-challenges-and-opportunities.

[159] Peter Desmond and Milcah Asamba, "Accelerating the Transition to a Circular Economy in Africa," in *Accelerating the Transition to a Circular Economy in Africa – Case Studies from Kenya and South Africa*, published May 8, 2019, accessed June 10, 2025, https://www.re-searchgate.net/publication/332416054_Accelerating_the_transi-tion_to_a_circular_economy_in_Africa.

[160] African Development Bank, "Africa Circular Economy Facility (ACEF)," accessed December 6, 2024, https://www.afdb.org/en/topics-and-sectors/topics/circular-economy/africa-circular-economy-facility-acef.

[161] Stats in the India section were all pulled from Dematdive, "Circular Economy," accessed December 12, 2024, https://dematdive.com/circu-lar-economy/.

[162] Doodlage, homepage, accessed December 6, 2024, https://doo-dlage.in/.

[163] National People's Congress of the People's Republic of China, *Circular Economy Promotion Law of the People's Republic of China* (2008), accessed December 6, 2024, https://ppp.worldbank.org/public-private-partner-ship/sites/ppp.worldbank.org/files/documents/China_CircularEcono-myLawEnglish.pdf.

[164] Simone Tam, "S Korea unveils strategy to promote circular econ-omy," Argus Media, June 22, 2023, accessed June 24, 2025, https://www.argusmedia.com/en/news-and-insights/latest-market-news/2461867-s-korea-unveils-strategy-to-promote-circular-economy.

[165] Asian Development Bank Institute, *Transitioning from a Linear to a Circular Economy: Case Studies from Asia and the Pacific* (2022), 12, sec. 1.2.1, accessed December 6, 2024, https://www.adb.org/sites/de-fault/files/publication/774936/adbi-transitioning-linear-circular-econ-omy-developing-asia-web.pdf.

[166] Asian Development Bank Institute, *Transitioning from a Linear to a Circular Economy*, 1.

[167] Asian Development Bank Institute, *Transitioning from a Linear to a Circular Economy*, 1.

[168] ASEAN, "ASEAN Adopts Framework for Circular Economy," October 21, 2021, accessed December 6, 2024, https://asean.org/asean-adopts-framework-for-circular-economy/.

[169] McKinsey & Company and The Business of Fashion, *The State of Fashion 2024* (2023), accessed December 6, 2024, https://www.mckinsey.com/industries/retail/our-insights/state-of-fashion-2024#/, 93.

[170] OEC, "Used Clothing," accessed June 23, 2025, https://oec.world/en/profile/hs/used-clothing.

[171] Joost van Barneveld et al., *Regulatory Barriers for the Circular Economy: Lessons from Ten Case Studies*, ed. Geert van der Veen (Technopolis Group, Fraunhofer ISI, thinkstep, and Wuppertal Institute, July 2016), accessed June 10, 2025, https://circulareconomy.europa.eu/platform/sites/default/files/2288_160713_regulary_barriers_for_the_circular_economy_accepted_hires_1.compressed.pdf.

[172] International Electrotechnical Commission, "IEC Technical Committee 111 (TC 111)," accessed June 10, 2025, https://tc111.iec.ch/.

[173] Boston Consulting Group and Platform for Accelerating the Circular Economy (PACE), *The Circularity Gap Report 2020: The Landscape of Metrics* (2020), accessed June 10, 2025, https://pacecircular.org/sites/default/files/2021-03/5faa4d272e1a82a1d9126772_20201029%20-%20BCG%20Metrics%20-%20White%20Papers%20-%20The%20Landscape%20-%20210_x_297_mm%20-%20bleed_3_mm.pdf.

[174] ReLondon, "Case Study: Toast Brewing – Leaving Breadcrumbs for Breweries to Follow," February 2, 2024, accessed December 17, 2024, https://relondon.gov.uk/resources/case-study-toast-brewing-leaving-breadcrumbs-for-breweries-to-follow.

[175] CGS, "CGS Survey Reveals Sustainability Is Driving Demand and Customer Loyalty," accessed December 17, 2024, https://www.cgsinc.com/en/infographics/cgs-survey-reveals-sustainability-is-driving-demand-and-customer-loyalty.

[176] Solitaire Townsend, "Consumers Want You to Help Them Make a Difference," *Forbes*, published November 21, 2018, updated December 10, 20201, accessed June 19, 2025, https://www.forbes.com/sites/solitairetownsend/2018/11/21/consumers-want-you-to-help-them-make-a-difference/?sh=504673e56954.

[177] NielsenIQ, "SPEND Z: Gen Z Spending Power and Influence," accessed December 17, 2024, https://nielseniq.com/global/en/landing-page/spend-z/

[178] First Insight, "The State of Consumer Spending: Gen Z Shoppers Demand Sustainable Retail," accessed

December 17, 2024, https://www.firstinsight.com/white-papers-posts/gen-z-shoppers-demand-sustainability.
[179] Krikke et al., "Circular Economic Surplus Asset Management," 122.

REUZEit
Next Gen Circularity